Afterwards

Folk and Fairy Tales
with Mathematical Ever Afters

Peggy Kaye

ETA/Cuisenaire®
500 Greenview Court
Vernon Hills, Illinois 6006-1862

Managing Editor: Doris Hirschhorn
Development Editor: Ellen Keller
Design Director: Phyllis Aycock
Cover Design and Illustration: Tracey Munz
Text design, line art, and production: Tracey Munz
Illustrations: Peggy Kaye

©Copyright ETA/Cuisenaire®
500 Greenview Court, Vernon Hills, Illinois 60061-1862

Table of Contents

Introduction 1

Diamonds and Toads 7

How Uncle Buzzard Lost His Feathers 22

Anansi and the Sky God's Daughter 34

Momotaro, the Peach Boy 46

Tiger, Brahman, and Jackal 60

Pecos Bill, the Greatest Cowboy of All 72

Cap O' Rushes 84

The Master Frog 96

The Queen Bee 109

About the Author

PEGGY KAYE is the author and illustrator of the *Games For* series: *Games For Math*, *Games For Reading*, *Games For Writing*, and *Games For Learning*. These books give parents and teachers hundreds of playful ways to help children master reading, writing, and arithmetic.

She is also the author of the *Homework: Math* and *Homework: Reading* series. These first- through sixth-grade workbooks encourage parent participation in their children's education. *Homework: Math* and *Homework: Reading* are now available in bookstores under the name *Enrichment Math* and *Enrichment Reading*.

In addition to books, she has written numerous articles and reviews for a wide range of magazines and newspapers.

Peggy Kaye lives in New York City where she writes, draws, tutors children, and works as an educational consultant.

Afterwards

Once upon a time, a teacher began telling fairy tales to her first- and second-grade students. Every Friday afternoon, she sat on a pillow-strewn floor, surrounded by eager youngsters, and began her story. Her students loved this Friday ritual. Even the roughest, toughest, most street-savvy kids fought to sit on the teacher's lap during tale time. And so the teacher discovered what she probably should have known all along—that fairy and folk tales touch something deep and abiding in children.

Years later, the teacher began reading about innovative educators who were using literature as a springboard for math lessons. Now, this teacher thought, why not hook up fairy tales and math? And so that is just what the teacher did.

Here is the result: a collection of nine tales and related math problems. The stories come from all over the world. The collection includes a tale from Nicaragua and another from Ghana. There are tales from India, Japan, England, France, Germany, Vietnam and a tall tale from the United States. The collection includes "how it came to be stories," a frog-prince fantasy, trickster tales, humorous stories, and yarns of love and bravery. In other words, there is something here for everyone.

Why is the book called *Afterwards*? Have you ever wondered what happens after "happily ever after"? If the people in a story are vivid enough, interesting enough, compelling enough, children delight in contemplating the characters' post-story lives. That's what they get with an *Afterwards* math story: an epilogue with a mathematical twist to a classic tale.

Imagine this. In the course of the French story, *Diamonds and Toads*, a kind girl gets the gift of unlimited jewels and flowers from a beneficent fairy. What will she do after the tale? Well, she plans to make a jeweled necklace as a gift for her special friend. Can you help her design one that follows a lovely color pattern? That's a math problem.

At the end of *Diamonds and Toads* the girl marries a prince and becomes, what else, a princess. In the *Afterwards* math story, the new princess offers flowers to everyone who visits the palace. She gives them away in a special manner. With a little counting or adding, you can find out how many flowers she gives to her loyal subjects. That's another math problem.

In crucial ways, *Afterwards* word problems differ from their basal cousins. As you will discover, the problems tend to be more thought provoking. They take longer to solve. They demand complex thinking skills. But the most important difference is, perhaps, a bit more subtle. Having heard the folk or fairy tale, children know the characters in *Afterwards* stories and have an authentic interest in their concerns, pleasures, and problems. And so these problems excite, rather than deaden, children's imaginations.

Using the Tales

There are nine tales in *Afterwards*. Each story is followed by several math problems. There is no established order for introducing the stories. Choose any one that seems appealing and begin. The same holds true for the math problems. Pick the one or ones you think are most appropriate for your students and ignore the rest. In general, it will take about 10 minutes to read the tale and another 20 to 30 minutes for children to solve each problem. You may decide to present more than one problem for any given tale. Do this on consecutive days or assign different groups of students their own problems.

Presenting the Tales and the Problems

Your first step is to share the stories with your class. You can read them, of course, but you might consider telling a story or two. Something special seems to happen when you tell rather than read a tale—maybe because these stories are rooted in an oral tradition. But it does take a little bit of work to master a tale before telling it. If you do want to put in the effort, don't try to memorize the tale. In general, storytellers prefer using their own words to create a spontaneous texture and tone. You can even transform the tale as you are talking, add to the plot or leave something out. Believe me, the Grimm brothers' ghosts will stay quiet in their graves.

Once you've presented a tale, turn to the math stories on the teacher pages. Select one that is suitable for your class. Then read it or tell it in your own words. Feel free to make any changes you want in the story or in the mathematical content. Use different numbers. Make the problem easier, or harder. Extend it. Condense it. Whatever you do is perfectly all right. On occasion, I suggest variations on problems, but even when I don't, vary away.

Many of the problems come with ready-to-duplicate student pages. If you don't want to photocopy them, you can always write the needed information on the chalkboard or a sheet of chart paper.

Grouping

Is it best to present the problems to your entire class or to smaller math groups? Should you have children work independently or in teams? There are no rules. The choice is yours. And you needn't feel tied down to one approach. Some problems will lend themselves more naturally to whole-class collaborations; others to small-group work, or to individual effort.

You do not need to present the problem immediately after reading the story. You can share a tale during a class meeting on Monday, present one problem to a math group on Tuesday, and another to a different group on Wednesday.

What should you do when a child or team of children finish before the rest of your class? There are several ways to handle this situation. You could ask the children to write a short explanation of their work. Even first graders can tackle this job if they use drawings and invented spelling in their responses. Alternatively, give the child or children a new *Afterwards* problem to solve. Better yet, encourage children to make up their own problems based on the *Afterwards* tale.

About the Problems

I've tried to include a variety of problems in *Afterwards*. There are addition problems and subtraction problems; graphing problems and geometry problems; counting problems and number-sense problems; estimation problems, pattern problems, and problems that focus on the base ten nature of our number system. The problems call on children to think logically, organize information, and employ various problem-solving strategies such as trial and error. As in life, some problems have only one answer; others have several possible solutions. The difficulty varies, too. This should make it easy for you to find just the right assignment for your students. But all the problems have one thing in common: They demand real thought and effort to solve.

What is a problem? Is finding the sum of 8 + 7 a problem? For you? Not really. It is more accurate to describe it as an unanswered question. A true problem demands concentrated intellectual energy to solve. It's never easy to unravel an authentic problem, and there is always a possibility of failure, no matter how hard you try. Proving Fermat's Theorem; well, now, that's a problem.

Afterwards is full of problems that first and second graders should find challenging (perhaps not as awesomely so as validating Fermat's Theorem, but challenging nevertheless). The problems will require time and effort to finish. At first, some children might have a hard time doing the necessary work to solve these problems. They may need considerable support from you. You must assure them that speed and praiseworthy work are not synonymous. Applaud youngsters' willingness to spend time exploring various ways to solve the problem. As children gain experience, the problem-solving process will become more natural. Children will feel confident and at ease. They will discover that intellectual intensity has a pleasure all its own.

Beginning a Problem

After reading a problem to your class or math group, spend some time discussing it. Make sure everyone understands what the problem asks. You might call on your students to restate the problem in their own words. Then elicit suggestions for solving the problem. If students fumble for ideas, offer one or two. Don't present a total strategy, just nudge your pupils in a promising direction. It helps if you ask questions: "I wonder if we should draw a picture of Momotaro's rice cakes?" Or "What would happen if we made a calendar to help Anansi work out his practice schedule?" Then let the children take over. Don't interfere if your students come up with less than promising proposals. Soon enough they will discover the pitfalls in their approaches and then they will have to rethink.

As they work, children may run into trouble. If that happens, feel free to help out. You can also encourage struggling children to talk to their classmates and discover how they are proceeding.

Good Strategies

When it comes to understanding a problem, nothing is more powerful than recreating the story for children. There are three excellent ways to do this.

- First, children can draw pictures showing, for instance, *Diamonds and Toads'* Princess Clarissa, the 12 visitors to the palace, and the proper number of flowers in each visitor's hand.
- Second, children can act out the problem. Let one child be the princess. She can dispense Snap™ Cubes instead of flowers. Then select children to approach Clarissa and receive their gifts.
- Third, recreate the story using manipulative materials. Make Snap™ Cube trains representing the flowers. The first train will have a single cube, the second will have two linked cubes, the third will have three, and so forth.

In general, you do not need to model the entire problem for your class. Just get pupils started. If you draw two or three stick figures clutching flowers and point out how the figures match the problem, your children will get the idea. Then they can take over the drawing and finish recreating the problem on their own.

To help understand and model each and every *Afterwards* problem, children will benefit from using manipulative materials such as Snap™ Cubes, Color Cubes, Unifix Cubes, kidney beans, or any other counting tool. This is especially true for harder problems. When tackling tough problems, your guidance and demonstrations of manipulative use are particularly important.

As you work the problems, you will begin to see changes in your students' problem-solving strategies. In a surprisingly short period of time, even without your prompting, students will begin grabbing counting materials, planning drawings, and acting out the math stories. When you expect great things from children, they meet the challenge.

Crossing the Curriculum

There are a variety of ways to link *Afterwards* stories to your language arts, social studies, and science programs. Here are a few thoughts.

Language Arts

- Children can write their own versions of an *Afterwards* tale. They can then illustrate their stories and "publish" them.
- Children might also write about the lives of characters after the story ends, only these stories do not need to have a mathematical hook.
- Encourage children to make up their own fairy tales based on the stories they hear in *Afterwards*.
- Read more tales to your class. There are so many good ones, you could easily make it a daily read-aloud event.
- Let children become storytellers by sharing their favorite tales with the class.
- Groups of children can act out a favorite tale. They don't need to write scripts. Let them improvise.

Social Studies

- Get a world map and stick it with pins indicating each story's country of origin.
- Cook ethnic foods from a selected *Afterwards* country.
- Find out about the music and art of the country.
- Do mini-research projects on *Afterwards* countries.

Science

- Several of the stories star animals. Have children research the life cycle of frogs, buzzards, or any of the other story creatures.

Of course, it isn't necessary to extend the tales. This is your program now, and you should feel free to use it in any manner you desire. In that way, I hope you will enjoy the *Afterwards* math lessons and that they all end happily ever after.

Diamonds and Toads

A Tale from France

Long ago and far away, a widow and her two daughters lived in a little cottage near the woods. The widow was a cruel, scheming, nasty woman, and so was Bettina, her older child. The younger daughter, Clarissa, was not anything at all like her mother or sister. Clarissa was kind, thoughtful, and very beautiful.

The mother loved Bettina, nasty though she was, and called her sweet names and gave her special treats. But this mother had no love at all for Clarissa. She never called Clarissa sweet names. Absolutely not. She called her Slop Girl. She never gave Clarissa treats. No indeed. Instead she made Clarissa work all day and half the night. "Cook the dinner, Slop Girl," she would say. "Wash the floors, Slop Girl. We need more water, Slop Girl."

Every day, rain or shine, Clarissa carried a large pitcher through the woods to a mountain stream. She filled the pitcher with cool, clear water and carried it back to the cottage.

One day Clarissa was returning from the stream when she saw an old woman dressed in rags sitting by the side of the road. As soon as this poor woman saw Clarissa, she called out, "Please, will you give me just one sip of your water? I am so thirsty."

"Of course," said Clarissa, "have all you want."

The old woman drank every drop of the water. When she finished, she said, "Forgive me, my dear, I did not mean to drink it all."

Clarissa smiled and replied gently, "There is nothing to forgive. I can easily get more water for my mother and sister."

"You are very kind," said the old woman. "Indeed, you are so kind that I want to give you a gift. From now on, whenever you speak, flowers and jewels will fall from your mouth."

With these words, the old woman left Clarissa. She hobbled off through the woods.

When Clarissa returned to the cottage, her mother was very angry. "Slop Girl, what took you so long?" she screamed. "Your poor sister is thirsty and weak from waiting for the water."

"I am sorry, Mother," said Clarissa. As she spoke, the most amazing thing happened. Three diamonds, real diamonds, and three pearls, real pearls, fell from her mouth.

Clarissa's mother could hardly believe her eyes. She grabbed the jewels and stuffed them in her pockets. "Slop Girl, how did you do that trick?" she demanded.

"The old woman promised me this gift," said Clarissa. "I never imagined her words would come true. She must have magic powers. Why, she must be a fairy from the fairy kingdom."

As Clarissa talked emeralds, tulips, rubies, and roses tumbled from her mouth.

"Tell me exactly what happened," said her mother.

After hearing Clarissa's story, the mother said, "Tomorrow your sister will go to the woods and meet this fairy. Yes, and then my dear Bettina will be richer than the duke, richer than the baron. Why, she will be richer than the king!"

The next morning Bettina headed for the mountain stream carrying a small jar in her arms. "Why should I tire myself out by hauling a huge pitcher? This little one will do," she said.

After a short walk through the woods, Bettina came to muddy creek. "I'll fill the jar here," she thought. "True, the water is dirty, but why should I keep walking? I'm not going to drink this."

After filling the jar, Bettina sat down on a rock and waited for the old woman to appear. A short time later, she heard someone behind her. Was it the old woman? No. It was a young lady walking through the woods. The young woman approached Bettina and said, "Please, I am very tired and so thirsty. Will you give me a sip of your water?"

Bettina frowned. "Why should I give you anything?" she snapped. "You are not the woman I am waiting for. Go and get your own water."

Bettina did not know it, but this was not an ordinary young lady. No, it wasn't. This was the fairy herself disguised as a young woman. Yes, it was.

No wonder then that Bettina did not worry when the young woman said, "You are thoughtless and cruel. For this, I will give you a gift. It is the gift you deserve. From now on, whenever you speak, toads, snakes, and other creatures will drop from your mouth." With these words, the young lady walked away through the woods.

Bettina did not worry one bit. The truth is, she was glad to see the young lady go. She did not want anyone else in the way when the old woman came along. But time passed and the old woman still had not come. Bettina got bored with waiting, so she returned to her home.

"Daughter, tell me, did you meet the fairy?" asked her mother as soon as Bettina reached the cottage.

Bettina started to talk, but when she opened her mouth, out popped four toads and three spiders. Bettina was shocked and disgusted.

So was her mother, who screamed, "This is your sister's fault. I will make her pay for hurting you!"

Her mother's angry words frightened Clarissa. She was so scared, she ran away from the cottage. She ran deep into the woods.

Now it so happened that the king's son, Prince Hal, was riding through the woods that day. As soon as he saw Clarissa running and weeping, he wanted to help.

"What are you doing alone in the woods?" asked Prince Hal. "Why are you running? Why are you crying?"

Clarissa told the prince everything. As she talked diamonds, emeralds, orchids, and marigolds fell from her mouth. Imagine the prince's surprise when he saw these treasures.

"Come with me to my father's castle," said Prince Hal. "You will be safe with my family."

By the time they arrived at the castle, Prince Hal and Clarissa had become good friends. The king and queen welcomed Clarissa into their home as a daughter. As time went by, Clarissa and Prince Hal decided to marry. So that's what they did, yes indeed. And they lived happily ever after.

Notes

About the Story

Diamonds and Toads is a French tale originally collected by Henri Pourrat. For my retelling, I referred to two different versions of the story.

Bjurström, C.G. ed. Tyler, Royall. translator. *French Folktales: From the Collection of Henri Pourrat*. New York: Pantheon Books, 1989, 54–58.

Opie, Iona and Peter. *The Classic Tales*. London: Oxford University Press, 1974, 98–102.

In addition to retelling the story, the Opies trace the tale's history and describe several similar stories. According to the Opies, you can find stories related to *Diamonds and Toads* in more than 20 countries. They even discovered one version that dates back to 1550.

Your Thoughts

Diamonds and Toads

The Flower Giveaway

In this problem, children count or add the consecutive numbers 1 through 12 in order to find the number of flowers that the princess gives away—78. As a follow-up, increase or decrease the number of people who visit Princess Clarissa and have the children again figure out the number of flowers Clarissa gives away.

Every day people line up at the castle to meet Princess Clarissa. And every day Clarissa talks with the first 12 people in line. As she talks, flowers fall from her mouth.

Clarissa likes to share these flowers with her visitors. She gives the first visitor one flower, the second two flowers, the third three flowers.

Clarissa keeps this up until she gives the twelfth visitor (the one who waited the longest time) 12 flowers. The prince asked Clarissa how many flowers she gave out each day. Clarissa could not figure out the answer. Can you help her?

Diamonds and Toads

Ten Flowers

In this problem, children investigate the number 10. As they make bouquets of red, yellow, and purple flowers, children have the opportunity to both count and add single-digit numbers.

Princess Clarissa decides to make bouquets with her flowers. She puts 10 flowers in each bouquet. One day she made a bouquet with 3 purple flowers, 3 yellow flowers, and 4 red flowers. She thought it was lovely.

Another day she made a bouquet with 2 purple flowers, 1 yellow flower, and 7 red flowers. For some reason, she did not like this bouquet very much.

Princess Clarissa did not worry, though. She knew she could make a new bouquet the very next day. You can help her.

Make bouquets for the princess. Use 10 flowers in each one. Make sure each bouquet is different.

Ten Flowers

Diamonds and Toads
A Fairy Necklace

This problem engages children in identifying and creating patterns. Although the worksheet shows only two necklaces, encourage children to make as many different necklaces as they want. Consider having children create different-length necklaces.

As you can imagine, Princess Clarissa is very grateful to her fairy friend. But how can she show her gratitude? The princess decides to make a jeweled necklace. She wants to give it to the fairy as a thank-you gift. The princess wants the colorful jewels in the necklace to follow a pattern.

Clarissa started working. The first necklace she made had this pattern: green, white, blue, green, white, blue, green, white, blue. The second necklace had this pattern: red, yellow, red, orange, red, yellow, red orange.

Clarissa is not satisfied with either necklace. Please design some more for her. Then choose your favorite. Princess Clarissa might agree with you.

Diamonds and Toads

A Fairy Necklace

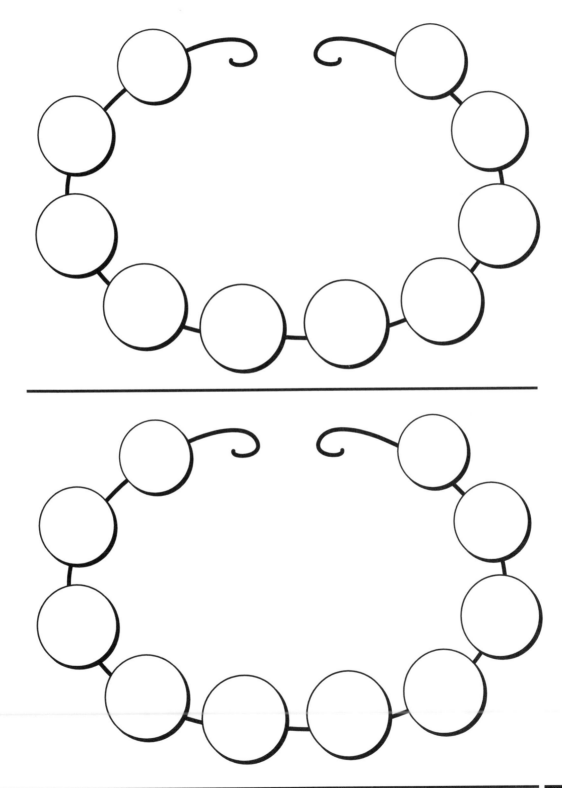

Diamonds and Toads
Disgusting or Terrific?

In this problem, children select one animal, survey a population, then represent their data in a bar graph. Children might canvas people in neighboring classrooms, the cafeteria, the school office, and the playground. Use the finished graphs to ask children questions such as these: How many opinions are recorded on the graph? Which column has the most (the least) opinions? How many more (fewer) people picked "terrific" than "sort of awful"? Did more or less than half the people pick "totally disgusting"?

Many people agree with Bettina, the mean sister. They think toads, snakes, lizards, and spiders are disgusting. Other people think differently. They think toads, snakes, lizards, and spiders are terrific.

Pick one of the creatures and find out how the people in your school feel about your choice. Asking people their opinion is called a survey.

Record your findings on a graph.

Diamonds and Toads
Disgusting or Terrific?

What do you think of _____**?**

Totally Disgusting	**Sort of Awful**	**Okay**	**Terrific!**

Diamonds and Toads

Jars of Jewels

To meet the challenge of this activity, children must find a way to divide a collection of jewels into four equal groups. The activity involves partitioning, counting, and one-to-one correspondence. Once children understand how to arrive at the solution (2 diamonds, 3 rubies, 6 emeralds, and 5 sapphires), create new problems by selecting a different multiple of 4 for each precious gem.

Last week these jewels fell out of Clarissa's mouth: 8 diamonds, 12 rubies, 24 emeralds, and 20 sapphires.

Clarissa wants to save all of her new jewels in 4 jars. She wants each jar to hold the same number of diamonds, rubies, emeralds, and sapphires.

How should she fill each jar?

Diamonds and Toads

Jars of Jewels

Clarissa's Jewels

8	diamonds
12	rubies
24	emeralds
20	sapphires

Diamonds and Toads

Just Right for a Witch

This multistep problem calls upon children to figure out different ways to represent 15 as the sum of five addends. It has five solutions: 1 box of spiders, 1 box of lizards, 3 boxes of snakes; 1 box of spiders, 2 boxes of toads, 2 boxes of snakes; 2 boxes of lizards, 1 box of toads, 2 boxes of snakes; 1 box of lizards, 3 boxes of toads, 1 box of snakes; 5 boxes of toads. Children are likely to use trial-and-error and "acting it out" as problem-solving strategies.

What about Bettina? What happens to her? Soon enough a neighboring witch hears about Bettina's talent for creating creepy, crawly, hoppy creatures. The witch needs these creatures for her spells, so she offers to buy boxes of creatures from Bettina. The witch insists that Bettina put only one type of creature in a box.

A full box will have 5 spiders, or 4 lizards, or 3 toads, or 2 snakes.

One day Bettina gives the witch 5 full boxes. The boxes contain 15 creatures all together. Can you figure out how Bettina filled the boxes?

Is there another way Bettina could fill 5 boxes with 15 creatures?

Just Right for a Witch

A full box will have 5 spiders.

A full box will have 4 lizards.

A full box will have 3 toads.

A full box will have 2 snakes.

How Uncle Buzzard Lost His Feathers

A Tale from Nicaragua

One day Uncle Rabbit was strolling on a country road when he came upon a patch of the most beautiful fruit trees and juiciest berry bushes imaginable. "What a treat," said Uncle Rabbit as he munched strawberries and gobbled peaches.

After a while, Uncle Rabbit saw the huge bird Uncle Buzzard flying through the sky.

"Hello, Uncle Buzzard, come and join me for a snack. This is absolutely and positively the best fruit I have ever eaten," called Rabbit.

Now Uncle Buzzard wasn't at all sure he wanted to join Uncle Rabbit. No, he was not. You see, Uncle Rabbit loved playing tricks. He especially enjoyed tricking Uncle Buzzard. Naturally, Uncle Buzzard did not like Rabbit's pranks. He certainly did not want to be tricked again today. Still, the fruit looked very tasty, and so Buzzard decided to take a chance.

He swooped down next to Uncle Rabbit and began nibbling blackberries. Sure enough, they were delicious. As he was eating, an idea occurred to Uncle Buzzard. What was the idea? It was a plan—a plan to trick Uncle Rabbit for once.

Uncle Buzzard giggled to himself as he plotted his trick. After thinking about every detail, he turned to Uncle Rabbit and said, "Yes, indeed, this is good fruit. It is not the best fruit, of course. You can only get the best fruit when you attend a feast in the clouds. You must agree, that fruit on the clouds is the ripest, the sweetest, the juiciest, the most perfect in the world."

"I have never eaten that fruit. I have never been to a such a feast. My, but I would like to go," sighed Rabbit.

"Well, why don't you?" asked Buzzard.

"Don't be silly, Uncle Buzzard. How can I get to the clouds? I can't fly," said Rabbit.

"What a shame, what a shame. Especially since there is a cloud feast this very afternoon. If only you could join us. You could bring your guitar and entertain the guests," said Uncle Buzzard.

"It is a shame," said Uncle Rabbit as he chewed on a bit of mango.

"Here's an idea," exclaimed Uncle Buzzard. "Let me fly you to the feast. You and your guitar can ride on my back. How about it? Do you want to fly with me?"

"That is a fine idea, and I gladly accept your invitation," said Rabbit.

"Go on, then. Get your guitar. I'll wait here for you," said Buzzard.

Uncle Rabbit was only gone a few minutes. When he returned, he climbed on Buzzard's back. Buzzard spread his wings and began flying. He flew higher and higher. When his wing touched the first cloud, Buzzard said, "Now, Rabbit, I will get back at you for all the nasty tricks you have played on me."

With that, Buzzard started flying in a crazy, mixed-up way. He swooped up and he charged down. He raced round and round.

What did Rabbit think about this? He was not happy at all. He yelled, "Buzzard, stop this wild flying! Stop right now!"

Very calmly Buzzard answered, "You don't like this way of flying, Rabbit? How sad. Well, then, here is something new."

Buzzard began to twist in loops and spirals. He zoomed backwards and rushed sideways.

"Buzzard," cried Uncle Rabbit, "this is even worse. Please, be a friend and stop this outrageous behavior!"

"It's only a bit of trick flying, Rabbit. But if you don't like it, I will do something new," laughed Buzzard.

Buzzard started zigging and zagging. He flipped and dipped. He twirled and whirled.

Uncle Rabbit was terrified. Somehow, some way, he had to stop Buzzard. Suddenly, he had an idea. He swung his guitar through the air and smashed it down on Buzzard. Uncle Buzzard's head was caught in the guitar! And then that great big bird fainted dead away.

When Uncle Buzzard fainted, he began crashing through the sky, heading to earth. Was Uncle Rabbit worried? No, not at bit. He knew just what to do. He grabbed Buzzard's wings and spread them out. Rabbit smiled as he floated slowly and gently down to earth.

When they landed, Buzzard woke up, but he could not see a thing because the guitar was still on his head. He shook himself up and down, back and forth. But the guitar did not move.

"Please, Uncle Rabbit, won't you pull your guitar off my head?" begged Buzzard.

"No, Buzzard, I will not. You wanted to trick me, but now the trick is on you. Come by my house later for some fruit pies, Uncle. That is, if you can get that guitar off your head," laughed Rabbit as he hopped away.

Uncle Buzzard pulled the guitar. He pushed it. He shoved it. Finally, he popped the guitar off his head. Do you know what happened as Buzzard pulled, pushed, shoved, and popped? He tore all the feathers off his head and neck. Yes, he did. And those feathers never grew back.

From that day to this, buzzards do not have head feathers or neck feathers. And from that day to this, buzzards know better than to try and trick rabbits.

Notes

About the Story

Before you tell the story, talk to your students about buzzards. Explain that they are large birds of prey, specifically vultures and condors. These birds have one very strange physical feature: They are featherless from the neck up. Perhaps you can track down pictures of condors or vultures to show your class. They are a sight worth seeing. After hearing the story, you might start a research project on these birds. If you do, you will undoubtedly find a more scientific explanation for buzzards' lack of neck and head feathers.

You can also use *How Uncle Buzzard Lost His Feathers* to introduce your class to "how it came to be" stories. There are hundreds of these tales. There are ancient myths—the Greek story of Persephone, for instance, tells how seasons came to be. There are modern tales such as the *Just So Stories* by Rudyard Kipling. These stories tell how the elephant got his trunk, and how the leopard got his spots. There is even an Iroquois version of buzzard and his missing feathers. You can find this tale in *Iroquois Stories: Heroes and Heroines, Monsters and Magic* by Joseph Bruchac, The Crossing Press, 1985.

To retell this story, I relied on one source.

Jagendorf, M. A. and Boggs, R. S. *The King of the Mountains: A Treasury of Latin American Folk Stories*. New York: Vanguard Press, 1960, 205–208.

Your Thoughts

How Uncle Buzzard Lost His Feathers
Fruit Pies

Children decide how many pies Uncle Rabbit will eat for
dinner, and then figure out how much fruit is needed
for that number of pies. To do this, children count, add,
and, possibly, skip count. Since the number of fruit pies
will vary, there will be many solutions to this problem.
When children's solutions are organized in a chart like
this one, patterns reveal themselves.

Number of Pies	Number of Apples	Number of Peaches	Number of Strawberries
1	2	5	10
2	4	10	20
6	12	30	60
12	24	60	120

"Fruit pies for dinner, fruit pies for dinner," sang Uncle
Rabbit as he hopped away from Uncle Buzzard.

On his way home, Rabbit stopped to gather fruit for
his pies. He knows that he needs 2 apples, 5 peaches, and
10 strawberries to make each pie.

But he does not know how many pies to make this evening.
And he does not know how much fruit he will need to prepare
the pies. Can you help Uncle Rabbit?

First, decide how many pies he should bake. Then figure out
how much fruit he should gather.

Remember, Uncle Rabbit is extremely hungry this evening.
After all, he had quite an adventurous day.

How Uncle Buzzard Lost His Feathers
Bragging

Children identify and then continue a growth pattern in order to find out how many people Uncle Rabbit brags to in a week. In the process, children do a great deal of counting and adding to find the solution—77 people.

Uncle Rabbit is a great trickster, it is true. But he is an even better bragger. Naturally, he cannot wait to tell everyone he meets about his triumph over Uncle Buzzard. He starts bragging early Monday morning. Then, every day for the next 7 days he boasts to more and more people.

On Monday, Uncle Rabbit tells 5 people about his adventures with Uncle Buzzard. On Tuesday, he brags to 7 people. On Wednesday, he tells 9 people and on Thursday, 11 people.

If he keeps up this pattern, how many people will Uncle Rabbit brag to by Sunday night?

Can you invent another bragging pattern for Uncle Rabbit?

If Rabbit follows your pattern, how many times will he brag in the next 7 days?

How Uncle Buzzard Lost His Feathers

Uncle Buzzard's Feather Collection

This problem draws children's attention to the base ten character of our number system. It leads children to count by 10s up to 150.

Uncle Buzzard misses his neck feathers. Oh, yes, he does. To cheer himself up, he starts a feather collection. Every day he collects exactly 10 feathers. Every day his collection gets bigger and bigger.

Can you figure out how many feathers he will have after 2 days?

Next find out how many feathers are in his collection after 3 days.

After that, find out how many feathers are in his collection after 5 days.

Then tell about his collection after 6 days and after 8 days.

How many feathers will he have after 10 days?

How many days will it take Uncle Buzzard to collect 150 feathers?

How Uncle Buzzard Lost His Feathers
Uncle Buzzard's Feather Collection

Uncle Buzzard collects 10 feathers every day.

How many feathers will he have after 2 days?

How many feathers will he have after 3 days?

How many feathers will he have after 5 days?

How many feathers will he have after 6 days?

How many feathers will he have after 8 days?

How many feathers will he have after 10 days?

How many days will it take Uncle Buzzard to collect 150 feathers?

How Uncle Buzzard Lost His Feathers
Uncle Rabbit's Job

To solve this challenging problem, children must determine combinations of bugs, seeds, and worms that earn $30, when each kind of item has a different value. As they search for solutions, children might count, add, multiply, and/or subtract. In fact, there are 19 ways to make combinations worth $30.

Uncle Rabbit's guitar is ruined and he wants a new one. There is only one problem—a new guitar costs $30. Uncle Rabbit must go to work in order to earn the money. Fortunately, his good friend Aunt Blue Jay owns the world-famous Happy Bird Restaurant.

Aunt Blue Jay and Uncle Rabbit work out a deal. Rabbit will collect worms, bugs, and seeds for the restaurant's cook. Aunt Blue Jay will pay him $1 for each jar of seeds he collects, $2 for each jar of bugs, and $5 for each jar of worms.

Uncle Rabbit knows one way to earn $30. He can sell Aunt Blue Jay 10 jars of seeds, 5 jars of bugs, and 2 jars of worms.

Uncle Rabbit knows there are other combinations of seeds, bugs, and worms that will equal $30, but he is not sure what they are. Can you help him think of more ways to earn the money he needs?

How Uncle Buzzard Lost His Feathers

Uncle Rabbit's Job

Uncle Rabbit earns:

    $1 for each jar of seeds.

    $2 for each jar of bugs.

    $5 for each jar of worms.

Uncle Rabbit can earn $30 if he sells

10 jars of seeds, 5 jars of bugs, and 2 jars of worms.

Think of more ways for Uncle Rabbit to earn $30.

How Uncle Buzzard Lost His Feathers
The Dart Game

This problem challenges children to find three-addend sums using the numbers 2, 3, and 5, (i.e., all 2s, two 2s and a 3, or two 3s and a 5), then see which sums differ by 9. The solution to the problem is 15 points for Rabbit and 6 points for Buzzard.

One day Uncle Rabbit was sitting in front of his burrow when he saw Uncle Buzzard floating overhead. Rabbit called out, "Hello, Buzzard! You're not still angry at me, are you? Come on, let's be friends again. Fly down and join me for a game of darts. What do you say?"

At first, Uncle Buzzard did not know what to answer. He did not trust Rabbit, no, not at all. But he figured he would rather have Rabbit as a friend than an enemy. And so, Uncle Buzzard decided to accept the invitation.

Uncle Rabbit and Uncle Buzzard started playing. First Rabbit threw three darts, and then Buzzard did. Unfortunately for Uncle Buzzard, he did not know how to add. That's why Buzzard had to believe Rabbit when he said, "I beat you Uncle Buzzard. I beat you by nine points."

Was Rabbit telling the truth? If he was, where did his darts land and what was his score? Where did Buzzard's darts land and what was his score?

The Dart Game

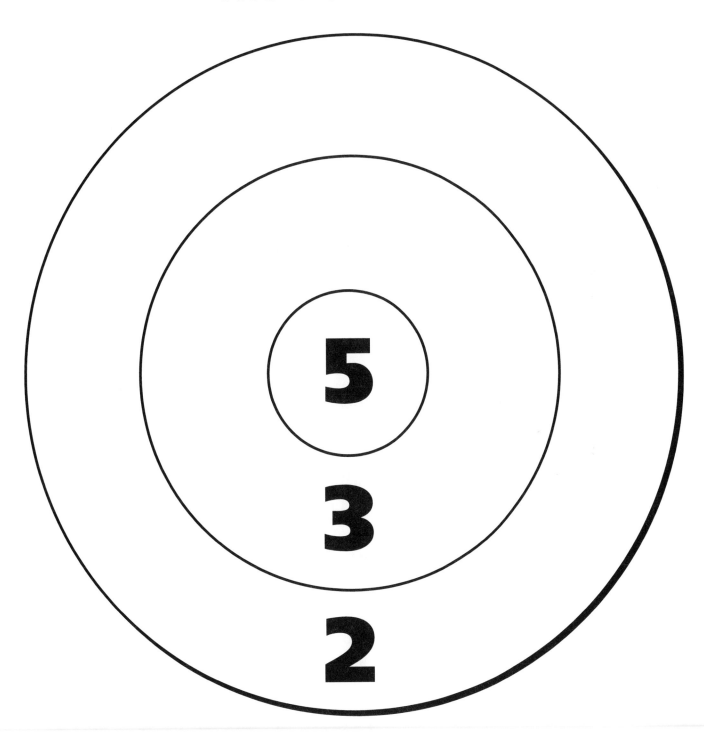

Anansi and the Sky God's Daughter

A Tale from Ghana

Everyone knows that Anansi the spider is the world's greatest trickster. Oh, the tricks he has played! Oh, the friends and enemies he has fooled! But once Anansi was too tricky for his own good. Anansi the trickster outtricked himself.

It started when the Sky God, Nyame (Nee-ah-mee), held a great feast. After dancing and eating, Nyame called everyone together. "I know that many of you want to marry my daughter," the Sky God said.

Tiger, Leopard, and all the others called out, "Yes, it is true."

"Listen to me," proclaimed Nyame. "I have decided that the first one smart enough to discover my daughter's secret name will be her husband."

Tiger, Leopard, and all the others listened carefully to Nyame's announcement. Then and there, they all decided, to try and learn the secret name.

The next day, Leopard walked to the Sky God's home. He had a list of 100 names. "One of these must be right," he said. But Leopard was disappointed. He did not know the secret name.

The sun set, the sun rose, and then Tiger went to Nyame. He, too, had a list of 100 names. He read everyone, but he never said the secret name. Next Hyena and Cheetah tried. Snake tried, too. They all failed.

Just like the others, Anansi wanted to discover the secret name. Unlike the others, he had a plan.

One afternoon Anansi climbed a mango tree in Nyame's garden. He waited there for Nyame's daughter and her servant to take their daily walk. When the two arrived, Anansi plucked a juicy mango from the tree and let it fall to the ground.

The servant saw the fruit drop. She lifted it up and said, "Oh, Beduasemanpensa (bed-wha-seh-mahn–pen-sah), look at this beautiful mango."

"Put it in your basket. Later we will take it to my father," said the Sky God's daughter.

As soon as the servant put the fruit in her basket, Anansi dropped another mango and then another and another. Each time the servant cried, "Oh, Beduasemanpensa, I found one more mango."

Finally the servant said," Beduasemanpensa, the basket is full. Come, we must return to your father's house."

When the two left the garden, Anansi climbed down from the tree.

"Beduasemanpensa, I know your secret name," Anansi laughed as he ran home.

When Anansi entered his house, he reached for his talking drum. Why? He wanted to teach his talking drum to say Beduasemanpensa. "Anyone can speak the name," Anansi said to himself. "But tomorrow I will show everyone just how smart I am. I will not speak the name. Oh, no. I will play it on my drum."

That night, Anansi was too delighted to sleep. He made up his mind to visit his good friend Abosom (ah-bo-sahm) the Lizard.

As soon as he entered Abosom's house Anansi cried out, "I know it! I know the secret name. Tomorrow I will go to Nyame and play it on my talking drum."

"Ah, Anansi, it is true. You are the smartest of all," said Lizard.

"And you are a good friend, Abosom," he replied. "Because of that, I will tell you the name. I am too excited to keep it a secret anymore."

For the rest of the night, Abosom sang the secret name while Anansi played it on his talking drum. When the sun rose, Anansi, with Abosom by his side, headed for the Sky God's home.

Leopard, Tiger, and everyone else in the village saw Anansi walking. "I wonder," said Tiger, "does Anansi know the secret name?"

"Let's find out! Let's find out!" everyone called out as they trailed Anansi to the Sky God's home.

When this parade reached its destination, Anansi bowed to the Sky God and said, "Nyame , I am proud to say that I know your daughter's secret name."

"If that is true," replied Nyame, "let me hear it."

Anansi started playing his drum. He played softly and then loudly. The Sky God listened.

"I do not understand your drumming, Anansi," said the Sky God.

"You must listen more carefully," replied Anansi. Then he played the drum again.

"I still do not understand," said Nyame.

"Abosom, tell Nyame what the drum says," cried Anansi.

"Beduasemanpensa is the secret name," sang Abosom.

"You are right, Abosom. That is the secret name. You are the first one to speak it. You will marry my daughter," announced Nyame.

Anansi and the Sky God's Daughter

When Anansi heard these words, he grew very angry. "You are not being fair, Nyame! I discovered the name. I played it on my drum," Anansi complained.

"You banged your drum, but I did not hear my child's name. Abosom said the name. He said it loudly. He said it clearly. He will marry my daughter," said the Sky God.

Everyone agreed. Abosom, the Lizard, should marry the Sky God's daughter.

"I guess the talking drum was one trick too many," said Anansi as he left Nyame's house alone.

Notes

About the Story

Anansi, that trickster spider, is a recent but beloved addition to the pantheon of folk-tale characters in the United States. Normally, Anansi, through clever manipulations, outsmarts all comers. This story breaks that mold. In this story, Anansi the trickster outtricks himself.

You may notice a resemblance between *Anansi and the Sky God's Daughter* and the European tale *Rumpelstiltskin*. In both stories, there is a quest to find a secret name. In fact, dozens of tales and myths center on power gained by finding a secret name. One of the earliest is an Egyptian myth in which the goddess Isis gains super powers by tricking the Sun God Ra into telling her his secret name.

You might use Anansi's story as a jumping off point for a unit on names. Where do names come from? What is the difference between a family name and a first name? Does your first name have a meaning? What is a nickname? What would you like for a secret name?

Anansi and the Sky God's Daughter is an African tale from Ghana. For my retelling, I referred to two versions of the story.

> Courlander, Harold. *The King's Drum and Other African Stories*. New York: Harcourt, Brace & World, 1962, 36–40.

> Jaffe, Nina. *Patakin: World Tales of Drums and Drummers*. New York: Henry Holt, 1994, 29–36.

In Nina Jaffe's excellent book, she explains something very interesting about Beduasemanpensa's name. "Bedua" means tenth born and in Ghana the tenth-born child is considered to be a very special person. Clearly, the Sky God would only let someone exceptional marry his tenth-born child.

I know that the name "Beduasemanpensa" is hard to pronounce. Say it whatever way makes you comfortable even if it differs somewhat from the provided guide. No matter how you say the name, the story will be interesting and fun for your children.

Your Thoughts

Anansi and the Sky God's Daughter
Loud and Soft

This problem engages children in finding permutations. A permutation is an arrangement of things in a definite order. Specifically, children arrange beats—loud and soft—to create eight different three-beat tunes, then 16 different four-beat tunes. The benefit of organizing data becomes apparent as children make certain that each tune they find is different from the rest.

Does Anansi give up his drumming because the Sky God's daughter marries Abosom? No, he does not. He drums every day. He can only play two notes on his drum: LOUD and soft. Still, he writes lots of songs. When he plays a two-beat song, he can create four different tunes:

LOUD LOUD

soft soft

soft LOUD

LOUD soft

Anansi can write 8 different tunes if he plays a three-beat song. Here are two of the tunes:

LOUD LOUD soft

LOUD LOUD LOUD

Can you find the other 6?

If Anansi uses four beats, he can play 16 different tunes. How many of the 16 can you find?

Anansi and the Sky God's Daughter

Practice, Practice, Practice

Allowed to use only the numbers 1, 2, 3, and 4, children must make 10 choices that add to 25. Children most commonly use trial-and-error, which gives them many opportunities to count and add. In time, children may realize that rearranging a solution or repeating combinations of numbers, like five 5s, creates new solutions without having to calculate.

Here are some examples.

Day	1	2	3	4	5	6	7	8	9	10
Hours	1	2	3	4	1	2	3	4	1	4
Hours	2	3	4	1	2	3	4	1	4	1
Hours	2	3	2	3	2	3	2	3	2	3
Hours	1	4	1	4	1	4	1	4	1	4

How can Anansi become a better drummer? He can practice, of course. And that is exactly what he decides to do. He plans to practice 25 hours in the next 10 days. He will practice 1, 2, 3, or 4 hours each day. He does not want to practice the same amount of time on any two days in a row. For some reason, he thinks this will bring him bad luck.

Make one practice schedule for Anansi to follow. Then see how many different schedules you can make.

Anansi and the Sky God's Daughter
Who Is Smarter?

This problem invites children to find more ways than Anansi and Abosom did to express 10 as the sum or difference of other numbers. Unlike typical addition and subtraction practice, here children get the answer and must work backwards to create the problem.

Who is smarter, Anansi or Abosom? Some think Anansi is smarter because he found the secret name. Others think Abosom is smarter because he married the Sky God's daughter.

After a lot of arguing, Anansi and Abosom plan a contest. They will each list ways to make 10 by adding and subtracting numbers. Everyone agrees that the one with the longest list must be the smartest.

Anansi and Abosom worked very hard on this problem. After a half an hour, they compared results.

Here is Anansi's list:

8 + 2	9 + 1
5 + 5	20 - 10
1 + 9	1 + 1 + 8

He found six ways to make 10.

Here is Abosom's list:

3 + 3 + 4	5 + 4 + 1
16 - 6	10 + 0
18 - 2 - 2 - 2 - 2	7 + 3

Abosom also found six ways.

It's a tie!

Can you beat Anansi and Abosom? Create your own list. How many different ways can you discover to make 10?

Anansi and the Sky God's Daughter
Who Is Smarter?

Here is Anansi's list:

8 + 2

5 + 5

1 + 9

9 + 1

20 - 10

1 + 1 + 8

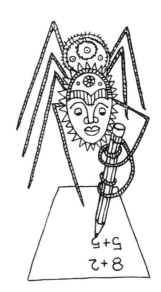

Here is Abosom's list:

3 + 3 + 4

16 – 6

18 - 2 - 2 - 2 - 2

5 + 4 + 1

10 + 0

7 + 3

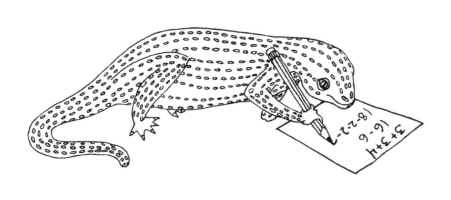

How many ways can you find to make 10?

Anansi and the Sky God's Daughter
The Wedding Feast

In an effort to help the cook know how many mangoes to pick from the garden, children figure out the number of mangoes each invited guest will eat. Although there is only one solution (94), this is a multistep problem involving one-to-many correspondence, counting, addition, and multiplication.

Nyame, the Sky God, wants to have a party to celebrate his daughter's wedding. Since the trees in his garden grow the best-tasting mangoes in the land, he decides to serve mango, and nothing but mango, at the celebration. There is only one difficulty—the cook does not know how many mangoes to pick.

The cook does know that 5 lizards, 3 tigers, 2 leopards, 10 rabbits, 5 cheetahs, 1 elephant, and 4 giraffes will attend the party.

The cook also knows that each rabbit eats 1 mango. Each lizard eats 2 mangoes. Each giraffe eats 6 mangoes. The elephant eats 10 mangoes. Tigers, leopards, and cheetahs eat 3 mangoes each. The Sky God eats 6 mangoes. The Sky God's daughter eats 4 mangoes.

But the cook is still confused. How many mangoes does she need? Can you help her find out?

Anansi and the Sky God's Daughter
The Wedding Feast

Guest List	
5 lizards	5 cheetahs
3 tigers	1 elephant
2 leopards	4 giraffes
10 rabbits	

Each rabbit eats 1 mango.

Each lizard eats 2 mangoes.

Each giraffe eats 6 mangoes.

The elephant eats 10 mangoes.

Tigers, leopards, and cheetahs eat 3 mangoes each.

The Sky God eats 6 mangoes.

The Sky God's daughter eats 4 mangoes.

How many mangoes does the cook need?

10 FOR THE ELEPHANT

Anansi's One-line Webs

Children engage in non-numerical thinking, finding solutions through experimentation. A discipline known as networking (the study of lines), however, deems it possible to predict when a figure can be drawn without lifting the pencil or retracing any lines. To do this, identify each vertex as even or odd. *Even* vertices are points in the figure where an even number of lines meet. *Odd* vertices are points where an odd number of lines meet. Figures with either all even vertices or exactly two odd vertices can be drawn from a single line.

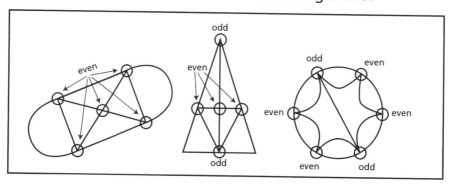

Anansi may have trouble drumming, but everyone agrees that he is the world's best web maker. On Abosom's wedding day, Anansi decides to show his enormous skill by decorating the wedding hall with amazing webs. Anansi has a special way of weaving. He always uses a single line of thread. Sometimes he crosses lines, but he never overlaps them. Anansi hopes his wedding-day webs will be so beautiful Nyame will wish he had selected Anansi to marry his daughter instead of Abosom the Lizard.

Take a look at three of Anansi's favorite webs. Can you copy them? Remember, you must use a single line and you must not overlap lines. After you copy Anansi's webs, try making some of your own one-line designs. Give them to a friend to copy.

Anansi and the Sky God's Daughter

Anansi's One-line Webs

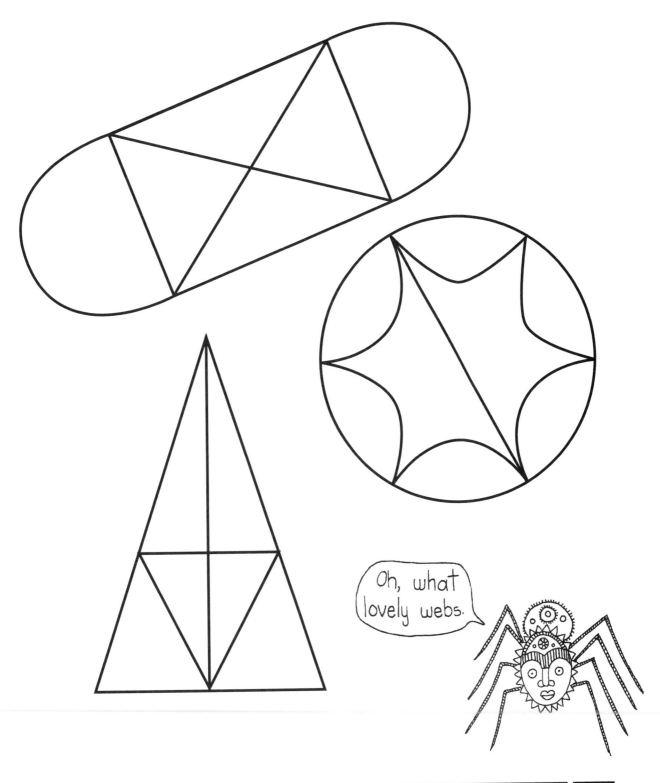

Oh, what lovely webs.

Momotaro, the Peach Boy

A Tale from Japan

Once upon a time, a kind old man and a gentle old woman lived in a small cottage near a large forest. The old man spent his days cutting bamboo trees. The old woman worked hard, too. They were poor people, but they were happy. Well, the truth is, they were not perfectly happy. They could not be perfectly happy because they did not have a child to love.

One day the old woman was washing clothes in a nearby stream when she saw a big peach, a huge peach. No, it was an enormous peach bobbing up and down in the water.

"I must catch that peach before it floats away," said the old woman. "I will give it to my husband for dinner. Oh, my, what a treat that will be."

The old woman tried to grab the peach, but every time her hands came near, the peach slid away. Then the woman had an idea. She started singing. Her song was so charming, the peach appeared to stop and listen. And then, odd as it may seem, the peach drifted toward the old woman. Indeed, it floated directly into her hands.

Carefully, the woman carried her treasure back to the cottage. When the old man came home, he said, "Wife, what have we here? This is the biggest, the hugest, the most enormous peach I have ever seen. Quick, get my knife. I will slice the fruit for dinner."

Just then the old man and the old woman heard a voice coming from inside the peach. The voice cried, "Please do not slice me. Please do not hurt me."

A moment later, the peach burst open and out crawled a beautiful baby boy. The baby reached his arms up and asked for a hug.

"Husband," said the old woman, "this child is surely meant for us. He is a gift from the gods. We will love him and care for him. We will call him Momotaro."

The old man and the old woman did care for Momotaro, and they loved him. Given such loving care, Momotaro grew up to be sweeter than a summer breeze and stronger than a hurricane.

You might imagine that now the old man and the old woman were perfectly happy, but this was not the case. You see, near the old couple's village, there was an island. On the island, there was a castle. In the castle, there lived an ogre. This ogre did not stay on his island. No, he did not. In the night, he raided neighboring towns and villages. He hurt people, stole money, and captured treasures.

The old man and the old woman did not have money or treasures. Even so, they were scared of the ogre. All the people in the land, rich and poor, were afraid of this fiend.

One day, when Momotaro was fifteen, he made a startling announcement. "Mother, Father, tomorrow I am going to Ogre Island. I plan to fight that monster. I will make sure he never hurts anyone again."

"Don't do this, Momotaro," cried the old woman. "You are strong, I know, but so is the ogre. His castle is surrounded by high walls and iron gates. He is protected by many cruel guards. Dozens of men have tried to defeat that villain, but each one died in the attempt. Please do not go. Stay home with your parents."

Momotaro's mother begged and pleaded, but she could not change her son's mind. So she decided to spend the rest of the night making rice cakes for his journey.

The next day, Momotaro started out on his way to find the ogre. After a time, he felt hungry, and so he sat down under a tree to enjoy a rice cake. At that moment, a dog, a big, huge, enormous dog jumped out from behind the tree.

"Give me that rice cake or I will take it from you," growled the dog.

Was Momotaro frightened? Not a bit. He turned to the dog and said, "Allow me to introduce myself. I am Momotaro, and I am on my way to defeat the ogre who lives on Ogre Island. I am not afraid of the ogre, and I am not afraid of you either. But if you want to be my friend, I will be happy to give you this rice cake as a gift."

"Excellent," replied Dog, "I will be your friend and I will accompany you on your journey. Working together, we will defeat the ogre."

After enjoying their rice cakes, Momotaro and Dog started walking through the forest. They had not gone more than a few feet when a monkey jumped out of a tree and landed in front of them.

"Hello, Momotaro," said Monkey. "I heard you and Dog talking. Now, give me a rice cake, and I will help you defeat the ogre."

"We don't need your help. Go away," snarled Dog.

At that, Monkey jumped on Dog and began swinging and punching. What did Dog do? He starting kicking and biting.

Momotaro stamped his foot. "Stop fighting!" he yelled. "Dog, don't you see, this is a brave monkey. I am sure he can help us."

With these words, Dog and Monkey stopped battling. They decided to be friends. Dog handed Monkey a rice cake, and the three friends started walking. They hadn't gone very far when a pheasant flew down at them from the top of a tree.

"Hello, Momotaro," said Pheasant. "I heard you talking to Dog and Monkey. Now, give me a rice cake and I will help you defeat the ogre."

"We don't need your help. Go away," said Monkey.

At that, Pheasant jumped on Monkey. Monkey and Pheasant started swinging and swatting and clawing.

Again, Momotaro stamped his foot. "Monkey!" he cried, "this pheasant is a brave fighter. We need brave fighters to battle the ogre."

When Monkey heard these words, he turned to Pheasant and said, "Momotaro speaks the truth. You are a good fighter. All right then, come with us to Ogre Island."

Now the four friends walked along together and soon they reached the sea, and far out in the middle, they saw Ogre Island. In no time at all, they were in a boat and on their way.

The ogre's castle was near the shore and was surrounded by huge stone walls. There was only one door through the walls, and it was protected by an iron gate. When Momotaro looked through the gate, he could see the guards marching up and down, back and forth.

Fortunately, Momotaro had a plan. He turned to Pheasant and said, "Here is a job for you, my friend. Fly over the wall. Attack the guards. Fight them. Scare them. Make them run."

Pheasant did not hesitate. He soared over the walls, saw the guards, swooped down, and began scratching, clawing, and pecking with his strong beak. Ogre's guards were terrified of that fighting bird and wanted to escape. In a panic, they pushed open the iron gate, ran out past Momotaro, Monkey, and Dog and swam off away from the island.

The ogre stood all alone in the courtyard. Momotaro came in punching. Monkey came in swinging. Dog came in snapping. The ogre got down on his knees and begged Momotaro to spare his life.

Momotaro stood looking down at the ogre. "I plan to let you live," he said, "but you will never again leave this island. Pheasant, Dog, and Monkey will see to that. From today on, they are masters of this castle. You, Ogre, will live in the dungeon."

Then Momotaro searched all the rooms in the castle. He discovered chests of silver, pots of gold, and jars of jewels.

Momotaro carried these treasures home to his parents. You can be sure that Momotaro's parents were now completely happy. Their son was home, they had all the money they could ever need, and never again did they have to worry about the ogre on Ogre Island.

Notes

About the Story

For this retelling I referred to two versions of *Momotaro.*

Uchida, Yosihiko. *The Dancing Kettle and Other Japanese Folk Tales.* New York: Creative Arts Book Company, 1986, 97–106.

Dolch, Edward and Dolch, Marguerite. *Stories from Japan.* Allen, TX: DLM Teaching Resources, 1960, 61–81.

Your Thoughts

Momotaro, the Peach Boy
Gold Blocks

To solve this problem, children must figure out how to separate bags of gold into two equal groups. Children count, add, and subtract, searching for the solution through a process of trial and error. The answer is 3 + 4 + 7 = 8 + 6.

Momotaro and his parents have more treasure than they need. Much more. What should they do with it all?

One day Momotaro's mother had an idea. "Let's share some gold with our neighbors," she said. "That is a wonderful plan," agreed Momotaro's father.

Momotaro thought so too. And so it was decided. But the family would not give gold away to just anyone. Oh, no. They would only give gold to a person who could solve a very tricky puzzle. Momotaro says anyone who can solve such a puzzle deserves a reward.

Momotaro called all his neighbors together. He showed them five bags. Then he filled each bag with a different number of gold blocks. He put 3 blocks in the first bag, 4 in the second, 6 in the third, 7 in the fourth, and 8 in the fifth.

Next Momotaro made a grand announcement. He said, "Who can divide these bags into two groups so that each group has the same amount of gold? The first one to do that will win all five bags."

What do you think? Is this a hard puzzle? Try it, and then decide.

Gold Blocks

Here are Momotaro's bags.
They are filled with gold blocks.

Put the bags in two groups.
Make sure each group has the same amount of gold.

Momotaro, the Peach Boy
The Dice Game

This is an addition game in which each child tries try to be the first to completely cover his or her game board. Along with practicing basic addition facts, children are introduced to probability: Some sums are more likely to occur then others.

Dog, Pheasant, and Monkey like living in the castle on Ogre Island. In the evening, they play a dice game to pass the time.

Here's how they play. First, everyone takes a game board and some beans, and picks six of these numbers: 2, 3, 4, 5, 6, 7, 8, 9, 10, 11, 12. Next, they write each of their numbers in a different section of their game boards.

Everyone agrees Dog can go first. Dog rolls two dice and adds the numbers together. He looks for this sum on his game board and puts a bean in that section. Then he gives the dice to the next player.

The players take turns rolling the dice, adding the numbers, and putting beans on their boards. The first one to put at least 1 bean in every section wins.

Dog thinks some numbers are luckier than others. Do you agree? If so, which numbers do you think are best? Why?

Momotaro, the Peach Boy
The Dice Game

Here is your game board.

Here is a number list: 2, 3, 4, 5, 6, 7, 8, 9, 10, 11, 12.

Pick 6 numbers.

Put one number in each section of your board.

Play the game.

Momotaro, the Peach Boy
The Rice Cakes

This is a logic problem that requires computation. To find out that Momotaro needs to bring 66 rice cakes to his friends, children engage in some combination of counting, addition, and subtraction.

Every few weeks, Momotaro visits his good friends on Ogre Island. He always brings a big pile of fresh rice cakes to share with Dog, Monkey, and Pheasant. He even brings some rice cakes for the ogre!

Can you figure out how many rice cakes he carries to the island?

Here are your clues:

- Pheasant eats 2 rice cakes.
- Dog eats 8 more than Pheasant.
- Monkey eats 10 more than Pheasant.
- Ogre eats 20 more than Monkey.
- Momotaro eats 2 less than Monkey.

Momotaro, the Peach Boy
The Rice Cakes

Momotaro knows that:

- Pheasant eats 2 rice cakes.
- Dog eats 8 more than Pheasant.
- Monkey eats 10 more than Pheasant.
- The ogre eats 20 more than Monkey.
- Momotaro eats 2 less than Monkey.

To feed everyone, how many rice cakes must Momotaro bring to Ogre Island?

Momotaro, the Peach Boy
How Many Peaches?

To help Momotaro decide how many peaches to pick, children must concentrate on the number of peaches remaining when each of several quantities is divided by 2, 3, and 5. The only solution that satisfies all the given clues is 27 peaches.

When Momotaro returned home, right after defeating the ogre, his mother was so delighted, she decided to bake her son a big, huge, enormous peach pie. Momotaro offered to pick peaches for the pie.

To tease her hero son, Momotaro's mother did not tell him exactly how many peaches to pick. Instead she gave him clues and challenged him to figure out the correct number.

Here are the clues:

- Momotaro must pick more than 20 peaches but less than 30.

- If he divides the peaches into 2 piles, there will be 1 extra peach.

- If he divides the peaches into 5 piles, there will be 2 extra peaches.

- If he divides the peaches into 3 piles, there will not be any extra peaches.

"If you can conquer an ogre," Momotaro's mother said, "you can certainly outsmart a tricky math problem."

But Momotaro is not sure he can solve the problem. Can you help him?

Momotaro, the Peach Boy
How Many Peaches?

Can you help Momotaro pick the perfect number of peaches?

Here are the clues:

- Momotaro must pick more than 20 peaches but less than 30.
- If he divides the peaches into 2 piles, there will be 1 extra peach.
- If he divides the peaches into 5 piles, there will be 2 extra peaches.
- If he divides the peaches into 3 piles, there will not be any extra peaches.

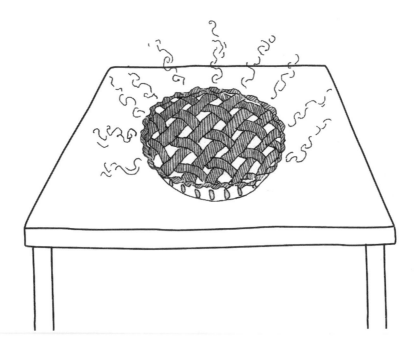

Momotaro, the Peach Boy
Shared Numbers

This problem, which focuses on sharing fairly, can contribute to children's understanding of the meaning of division. As they find the quantities that Dog, Pheasant, and Monkey can share equally with nothing left over, children identify the multiples of 3 between 1 and 100.

Dog, Pheasant, and Monkey enjoy living together. Soon, though, they realize that they must share absolutely, yes, absolutely everything. They must share food, clothes, and games. And they must share evenly or they will begin fighting one another.

They know they can share 6 peaches, 6 rice cakes, or 6 books evenly. They know they cannot share 8 peaches, 8 rice cakes, or 8 books evenly.

But, they fight about the other numbers.

Help these friends stop fighting. Find all the three-way share numbers you can between 1 and 100.

Momotaro, the Peach Boy
Shared Numbers

1	2	3	4	5	6	7	8	9	10
11	12	13	14	15	16	17	18	19	20
21	22	23	24	25	26	27	28	29	30
31	32	33	34	35	36	37	38	39	40
41	42	43	44	45	46	47	48	49	50
51	52	53	54	55	56	57	58	59	60
61	62	63	64	65	66	67	68	69	70
71	72	73	74	75	76	77	78	79	80
81	82	83	84	85	86	87	88	89	90
91	92	93	94	95	96	97	98	99	100

Tiger, Brahman, and Jackal

A Tale from India

One sunny day, a Brahman, a very religious man, was walking through the woods when suddenly he saw Tiger. Was the Brahman scared? No, he was not. You see, Tiger was trapped in a cage.

When Tiger caught sight of the Brahman, the caged beast cried out for help. "Oh great one," he sobbed, "please release me from this cage. If you do not, my fate is sealed. The hunters will return, and they will kill me. Have pity, honorable one. Open the cage door and save my life."

The Brahman heard Tiger's words, and he felt great sympathy for the creature. But he did not want to let him out of the cage. Can you guess why? And so the Brahman said, "Tiger, poor Tiger, I do feel sorry for you. Yes, very sorry indeed. But I cannot let you out of that cage. If I do, you will eat me."

"I will not eat you," protested Tiger, "absolutely not. I will be grateful to you. I will take care of you and protect you forever and a day."

As Tiger talked, tears streamed down his face. He was so miserable, so pathetic, that the Brahman's heart melted. "Stop crying, friend Tiger," said the Brahman. "I will do as you ask. I will set you free."

As soon as the Brahman opened the door, Tiger leaped out of the cage and jumped on top of the Brahman.

"You are a foolish man to trust the words of an imprisoned animal," said Tiger. "Now everything is different. You are in my trap. I wonder, what shall I do with you? I guess I will eat you for dinner. After spending so much time cooped up in that cage, I am exceedingly hungry."

In this awful situation, what did the Brahman do? He pleaded for his life. "Tiger," he begged, "remember your promise. You said you would be grateful to me, grateful forever. I am not asking you to take care of me. Certainly not. I am simply asking you not to eat me."

"Did you really think I would be grateful?" laughed Tiger. "No one is ever grateful. But, to show that I am a fair-minded animal, I will give you a chance to save your life. Find one person or one thing who believes in gratitude. Find one person or one thing who thinks gobbling you up is unjust, and I will let you go."

Quickly the Brahman looked around for someone or something who would say, "Tiger, you must free this worthy man. It is not right to take his life."

Then he saw a pipal tree. "Pipal Tree, kind Pipal Tree," he said, "you saw what happened. You saw me free Tiger. You must agree, he should show proper gratitude and let me live."

The pipal tree answered in a cold, angry voice, "Gratitude? Don't talk to me about gratitude. For years and years, here I stand giving shade and shelter to one and all. Does anyone show me gratitude? No, they do not. Instead they pull off my branches and strip my leaves to feed their cattle. You must face your fate, Brahman. I can do nothing for you."

The Brahman was not happy to hear these words. No, he was not. He did not give up hope, though. Nearby he saw a water buffalo. The buffalo was tied to a huge wheel. He was pulling the wheel round and round.

"Buffalo, good , strong buffalo, you must believe in gratitude. Tell Tiger to do the right thing and set me free."

The buffalo laughed at these words. "Gratitude?" he snickered. "Only fools expect gratitude. Listen to my story. I lived with a family. I gave them milk every day. And while I did, they petted me; they loved me; they fed me delicious seed cakes. Now I have no more milk to give. Does that family show gratitude for all my past services? I should say not. Instead, every morning they tie me to this wheel. I work and work all day. No one pets me; no one loves me; no one gives me good food to eat. Gratitude, ha!"

Now the Brahman did not know to whom he could turn, and so, in desperation, he looked down and addressed the road. "Road," he said, "you must give your opinion."

"Every day, all day, old people and young people, large people and small people, dogs, pigs, cows, and so many others trample and stamp on me. And do they ever show gratitude? I should say not. They throw trash on me, they spit on me, and kick up my stones. Don't ask me about gratitude" grumbled the road.

Hearing these words, Tiger laughed, but the Brahman was so scared and so unhappy, he began to cry. Just then Jackal came strolling along. He saw the Brahman weeping and he asked, "Tell me sir, why are you crying?"

Through his tears, the Brahman told his story. But Jackal did not seem to understand.

"How confusing," he said. "Can you explain what happened again?"

"Listen to me," said Tiger. "I will explain it all to you."

Tiger retold the story, but Jackal was still befuddled. "Let's see," he said, "our friend the Brahman was in the cage and then..."

"No, no," Tiger interrupted. "I was in the cage."

"I was in the cage," repeated Jackal solemnly. "But, Tiger, I was not in the cage. That cannot be right."

"Of course it's not," shouted Tiger. "Listen and I will explain again."

Tiger started telling the tale. After a moment, Jackal burst out, "I see it all now. Tiger was in the Brahman. No, no, that cannot be right. Was the Brahman in Tiger? My, this story is so confusing, I'll never understand it."

"Oh yes you will," growled Tiger. "I will make you understand. I don't care how long it takes!"

"Thank-you, Tiger," said Jackal. "I promise to listen with great care to your words."

Then Tiger began again, "I was in the cage. That means me, Tiger. Just Tiger and only Tiger. I was in the cage!"

"In the cage," said Jackal. "I see. But Tiger, how did you get in the cage?"

"I got in like this," shouted the angry Tiger as he stamped into the cage. "Now, you silly animal, do you understand?"

"Yes, I understand perfectly," said Jackal. And with these words, he slammed the cage door, locking Tiger inside.

Tiger cried and begged, but this time, no one let Tiger out of the cage.

Now you know how Tiger learned the price of ingratitude and how the Brahman, with Jackal's help, lived happily ever after.

Notes

About the Story

To understand this story, your students must know about Brahmans, jackals, and pipal trees.

What is a Brahman? A *Brahman* is a priest. There are four classes in traditional Indian society. The Brahmans are the most respected and highest ranked of the four.

What are jackals? *Jackals* are doglike animals native to Africa and southern Asia. They eat plants and small animals.

What is a pipal tree? A *pipal*, also known as a bo, is a gigantic tree native to Asia. It is said that Buddha sat under a pipal tree at the moment of his enlightenment. Therefore, artists use pipal-tree renderings as symbols for spiritual illumination. In Sri Lanka there is a living pipal that many believe started life as a cutting from Buddha's actual enlightenment tree.

For this retelling, I relied on two sources.

Jacobs, Joseph. *Indian Fairy Tales*. New York: Dover Publications, 1969, 66-69.

Dolch, Edward and Marguerite, Jackson, Beulah. *Far East Stories: For Pleasure Reading*. Allen, TX: DLM Teaching Resources, 1953, 127–135.

The Dover publication is an exact reproduction of the 1892 edition of Jacobs' work. You can also find the Jacobs' version of this story in a second book:

Haviland, Virginia. *Favorite Fairy Tales Told in India*. New York: Beech Tree Paperback Books, 1973, 81–87.

Your Thoughts

Tiger, Brahman, and Jackal
What's the Difference?

This game provides addition and subtraction practice. To help children work independently, show them how to make Snap Cubes trains and then find the difference between them. Pick two numbers, like 8 and 5. Make an 8-car train and a 5-car train. Then, place the trains side-by-side. Snap off the "extra" cubes and count them. Explain to children that the number of "extra" cubes is the difference between the two numbers.

The Brahman was very grateful to Jackal. After all, this smart animal saved his life. So the Brahman invited Jackal to live in his home. Jackal was happy to accept. Every night after dinner, the Brahman and Jackal played a game—a card game called "What's the Difference?"

You can play too. Here's how.

First, find eight cards, each 3 inches by 5 inches. Cut the cards in half. Then write the numbers 0 through 15 on the cards. Write one number on each card.

Now find someone to play with. Place the cards face down on a table.

Take turns. On your turn, pick two cards. One number will be bigger than the other. Find out how much bigger and record this amount. For example, if you pick a 9 and a 4, you would write 5 on your score sheet.

On your next turn, again pick two cards. Find the difference between the numbers. Add the difference to your score. Continue doing this until one player reaches 50.

The first player to earn 50 points wins.

Tiger, Brahman, and Jackal
What's the Difference?

Place the cards face down.

Take turns. Pick two cards. One number will be higher. How much higher? Find out.

That number is your score.

Keep going until one player gets 50 points.

_____	_____
PLAYER	**PLAYER**

Tiger, Brahman, and Jackal

The Thank-You Contest

Children should discover that Jackal says 100 thank-yous whereas Brahman says 110. This challenging multistep problem engages children in adding, counting by twos, and counting by tens.

Once day Jackal and the Brahman were talking about their adventure with Tiger.

"I really learned a lesson," said the Brahman. "From now on, I will say thank-you and thank-you and more thank-yous than you can imagine."

"That is a good idea," said Jackal. "But I bet I can say more thank-yous than you!"

"All right," said the Brahman, "let's have a contest. Let's see who can say the most thank-yous in the next 10 days."

"Fine," said Jackal, "for the next 10 days, I will say 10 thank-yous each day."

"I have a different plan, a pattern plan," said the Brahman. "I will say a new number of thank-yous every day. I will start with 2, and the next day I will say 4, and the next day I will say 6. I will keep up my pattern for 10 days."

If they follow their plans, who will win the thank-you contest?

Invent your own thank-you plan.

If you follow your plan, how many thank-yous will you say in 10 days?

Thank-you, Thank-you, Thank-you

This series of "thank-you" problems offers children opportunities to add, count, skip count, look for patterns, estimate, and organize and interpret data. Some solutions are dependent on class size or on how fast a child can write. The solutions to the problems about classroom objects are multiples of 10.

The Brahman learned to thank everyone and everything. He felt that even trees and roads need to hear a thank-you sometimes.

Here are some thank-you problems to solve:

- If you say thank-you to everyone in class, how many thank-yous will you say?

- If you say thank-you to everyone in class every day from Monday until Friday, how many thank-yous will you say?

- If you say thank-you twice to every boy and three times to every girl, how many thank-yous will you say? What happens if you say thank-you three times to each boy and two times to each girl?

- Name 10 things in your classroom, like a book, a chair, your desk. If you say 2 thank-yous to each one, how many thank-yous will that be? If you say 3 thank-yous to each one, how many will that be? Try saying 4 thank-yous to each one. Look for a pattern. Use it to predict the results of saying 8 thank-yous to everything.

- If you write the word "thank-you" 10 times, how long do you think it will take? Find out.

Tiger, Brahman, and Jackal
A Number Trick

To do this trick, children must add and subtract several times. Before distributing copies of page 69, set the stage. Ask each child to write down a secret number between 5 and 20. Read each step of the trick aloud, allowing time for children to do the arithmetic. Have volunteers, one at a time, tell their results. Silently add 5 to each result and announce the sum as the secret number.

What happened to Tiger? The hunters did find him in the trap, but they did not kill him. No, not at all. You see, these hunters worked for the circus. They needed a wild animal for the circus show. So Tiger joined up.

Six months later, Jackal received a letter from Tiger. It said:

Dear Jackal,
   Thank you for tricking me back into that cage. Now I have a wonderful life. I am the star of the circus. When I enter the center ring, the audience stands up and cheers. My trainer is kind and generous. I live in a lovely home and get plenty of delicious food.
   Of course, I have not forgotten the importance of showing gratitude, and so I am sending you a thank-you gift. Actually, the gift is from my trainer. When I told him about how you tricked me, he laughed and laughed. Then he said, "Please share one of my favorite number tricks with your clever friend." Here is the trick. I hope you like it.

Sincerely yours,
Tiger

THE TRICK
Think of any number between 5 and 20.
Add 5 to your number.
Subtract 2 from your new number.
Now, add 8.
Next, subtract 10.
Then, add 4.
Finally, subtract 10 again.
Tell me how much you have now,
and I will tell you your secret number.

Tiger, Brahman, and Jackal
A Number Trick

Try Tiger's trick on a friend.

Give the friend these directions.
- Think of a secret number. It must be between 5 and 20.

- Add 5 to your number.
  But don't tell me your answer.

- Now subtract 2.

- Add 8.

- Subtract 10.

- Add 4.

- Subtract 10 again.

Now ask your friend to tell you that answer.

Add 5 to it.

Don't tell your friend what you are doing.

When you are ready, announce your answer.
This is your friend's secret number.

Tiger, Brahman, and Jackal
Gifts for All

Trial-and-error is a good way to solve this problem. In doing so, children have the opportunity to double numbers, do addition, and refine their guesses. The Brahman's lucky number is 3.

In his happiness, the Brahman did not forget the pipal tree, the water buffalo, or the road. Indeed he decided to give gifts to each one. "It will make them feel appreciated and cared for," he said to himself.

One day, the Brahman took gifts into the woods. He had brightly colored ribbons to tie on the pipal tree's branches, freshly baked seed cakes to feed the water buffalo, and trash baskets to line the sides of the road. Now the road would be trash free.

All together he had 21 gifts. Can you figure out how many trash baskets, seed cakes, and ribbons the Brahman gave away?

Here are your clues:

- The Brahman has a lucky number. It is less than 5. That is the number of trash baskets he gave to the road.

- If you double the Brahman's lucky number, you will know how many seed cakes he gave the buffalo.

- If you double the number of seed cakes, you will know how many ribbons the Brahman gave the pipal tree.

What is your lucky number? If the Brahman used your lucky number for the trash baskets and then doubled it for the seed cakes and doubled again for the ribbons, how many gifts would he give away?

Tiger, Brahman, and Jackal
Gifts for All

The Brahman gives 21 gifts in all.

- The number of trash baskets he gives to the road is the same as his lucky number.
  It is less than five.

- The number of seed cakes he gives to the buffalo is the same as his lucky number doubled.

- The number of ribbons he gives to the tree is the same as the number of seed cakes doubled.

What is the Brahman's lucky number?

How many trash baskets does the Brahman give away?
How many seed cakes does he give away?
How many ribbons does he give away?

Now imagine that Brahman uses your lucky number.
How many trash baskets will he give away?
How many seed cakes will he give away?
How many ribbons will he give away?
How many gifts will he give away?

Pecos Bill, the Greatest Cowboy of All

A Tale from the United States

Pecos Bill was born in Texas many, many years ago. He was an amazing baby. Oh yes, he was. He did not drink regular milk. No, indeed. His Ma gave him milk from a mountain lion because that's all baby Pecos would swallow. He would not play with his 10 brothers and 10 sisters, not Pecos Bill. So his Pa trapped a grizzly bear and brought the critter home to his baby boy. Baby Pecos wrestled that bear, and the bear wrestled back, and they had a fine time and became great friends.

When Pecos was one year old, his Ma and Pa decided to leave their rundown ranch and find a new home. Pecos, his brothers, sisters, and Ma and Pa crammed into a wagon, and the family started traveling. Soon they reached the Pecos River. That's when something awful happened. Baby Pecos fell out of wagon and into the water. The river carried Pecos away so far and so fast, that his Ma and Pa could not find their baby even though they looked and looked.

Pecos floated down the river for a week before landing on shore. Boy, oh boy, was he hungry! Near the riverbank, he saw a pack of coyotes feasting on deer meat. It seems those coyotes had never seen a human baby before. They thought Pecos was some new type of coyote pup. So they gave him a big hunk of raw meat. Then they adopted him into their pack. Pecos spent 10 happy years running with the coyotes. The coyotes taught him all their tricks. After a time, Pecos was the fastest runner, the loudest howler, and the best hunter of any coyote in the land.

One fine night, a cowboy named Tom came riding over the Texas hills. You can bet Tom was surprised to see a boy, a human boy, with no clothes on, sitting by a rock and howling at the moon.

"Hello," said Tom. "What are you doing way out here?"

Now Pecos did not understand human language, so he snarled and growled and prepared to attack. But Tom just smiled and hummed a tune. Pecos liked Tom's song, and he lay down to listen. Then Tom offered Pecos a slice of cooked meat. Pecos had never eaten cooked meat before. Deary me, he liked that supper better than any coyote meal.

Pecos and Tom spent the next three days together. In that time, Pecos learned to talk human talk. Then Tom said, "You better leave your coyote friends and come be a cowboy with me."

"I can't do that," Pecos said. "I am a coyote. Listen to me howl. Look at my fleas."

"Well," said Tom, "any good cowboy can howl, and all cowboys have fleas. But, to be a coyote, you've got to have a tail. Now, Pecos, look for yourself. You do not have a tail."

Pecos twisted left and turned right. No tail. "I guess I'm not a coyote after all," mumbled Pecos. Then and there, he decided to leave the wilderness and become a cowboy.

Tom gave Pecos some old clothes and the two set off over the hills with Tom riding his horse and Pecos walking alongside.

Soon Pecos got tired of walking. Just then he saw a huge mountain lion sleeping in a cave. "Wake up, lion," shouted Pecos.

That lion did wake up and he was in a mean mood. He went after Pecos with claws set to slash and teeth ready to rip. Was Pecos scared? Of course not. He picked up that angry beast and held him high in the air. Then Pecos tickled the lion under his chin. This was most agreeable to the lion, and he started purring like a kitten.

"Now, Lion, let's agree to be friends," said Pecos. "As your friend, I promise to give you lots of chin tickling. As my friend, you promise to let me ride on your back. Together we will go here, there, and everywhere." And that's just what they did.

Soon Pecos and Tom joined up with a crew of cowboys. Well, Pecos liked living with the cowboys, and he began inventing things to improve life on the ranch. Did you know that it was Pecos Bill who invented the lasso? Now, that's a story. You see, one afternoon when Pecos was riding his mountain lion around the Texas hills, he found himself looking down on the biggest rattlesnake that ever rattled. Was Pecos worried? Not a bit. He jumped off his lion and looked that hissing snake right in the eye.

"Now Rattlesnake," Pecos said, "it seems you want to fight. Well, okay, then. But let's make it a fair fight. You go right ahead and take three bites out of me. Make them your best because when you are done, I'll start fighting back. And Mr. Rattle, I plan to win this battle. When I do, I will drain out all your poison."

So that rattler bit Pecos once, twice, three times. Did Pecos mind? Absolutely not. Why, he smiled. Then he scooped up that snake and squeezed half as hard as he could. Straightaway, all the poison dripped and dropped out of that snake.

"Oh no, oh no," groaned the snake, "what have you done to me? Without poison, I cannot protect myself."

"Don't you worry, Mr. Rattle. You can live with me now. I'll protect you, and we will be great friends."

Then Pecos wrapped the snake around his neck as though it was a scarf. But before Pecos could climb on his mountain lion, what do you think he saw? He saw a gigantic lizard dashing through the sagebrush.

"Well friend Snake," said Pecos, "I do believe you can help me catch that big old lizard."

Pecos looped the rattler into the flying knot we call a lasso. He swung the lasso around the lizard's neck. The lizard was stuck—caught by the first lasso that ever lassoed.

"It works," shouted Pecos. "This is my best invention ever!" Pecos was so delighted he went out and lassoed five lizards and six cows before sunset. That evening, he showed all the cowboys on the ranch how to lasso. Of course, the cowboys used rope, not rattlesnakes.

After a time, Pecos thought he should get married and have children. Did this unusual man marry some ordinary woman? No, indeed. Pecos married Slue-foot Sue, the strongest cowgirl in the west. Sue was so powerful she rode up and down the Pecos River on top of a catfish. And that catfish was the size of a blue whale. When Pecos saw her ripping downstream, he shouted, "You are the woman for me!" Sue agreed and they got married.

Pecos and Sue lived together for many happy years. They had a big family, too. They raised boys and girls, but that's not all. They adopted stray coyote pups, mountain lion cubs, and bear babies. And that is how they lived happily ever after.

Notes

About the Story

Pecos Bill is an American tall tale. Tall tales are part story and part bragging contest. They also serve as "how it came to be" tales. In Pecos Bill, for instance, you discover the origin of that essential piece of cowboy equipment, the lasso. Other versions of *Pecos Bill* explain the forming of the grand canyon and the history of cowboy songs.

The form of a tall tale is different from a classic fairy tale or folk tale. There isn't a plot in the traditional sense. Instead, the lead character goes from event to event performing magnificent deeds. The humor and outlandishness of these stories explain their special appeal to children.

You might want to make sure your students know what a lasso is before beginning the story.

For this retelling, I used three versions of the tale.

Botkin, B. A., ed. *A Treasury of American Folklore: The Stories, Legends, Tall Tales, Traditions, Ballads and Songs of the American People*. New York: Crown, 1944, 180–185.

Marcatante, John J. and Potter, Robert R. *American Folklore and Legends*. New York: Globe Book Company, Inc., 1967, 201–207.

Stoutenburg, Adrien. *American Tall Tales*. New York: Viking Press, 1966, 24–36.

Your Thoughts

Hats and Boots

As they sort the information given in this problem, children must deal with one-to-one correspondence as well as many-to-one correspondence. To discover that Sue must make 22 hats and 64 boots, children add, count, and possibly skip count. Many draw pictures or "act it out." Writing a letter helps children focus on their problem-solving processes.

Pecos Bill never goes any place without wearing his cowboy hat and his cowboy boots. (Did you know it was Pecos who invented spurs for cowboy boots? Oh, yes it was.) Slue-foot Sue, being a true cowgirl, has her own hat and boots. But what about their children and animal babies? Sue wants them to have cowboy hats and boots, too. She decides to make hats and boots for everyone. There is only one problem. She doesn't know how many hats and boots to make.

Pecos and Sue have 6 sons, 6 daughters, 5 coyote pups, 3 mountain lion cubs, and 2 bear cubs. Sue does know that her children need 1 hat and 2 boots each. She knows that coyotes, mountain lions, and bears need 1 hat and 4 boots each.

Help Sue solve her problem. Then write her a letter explaining how you got the answer.

Pecos Bill, the Greatest Cowboy of All
Hats and Boots

Sue wants to make hats and boots for her:

- 6 sons
- 6 daughters
- 5 coyote pups
- 3 mountain lion cubs
- 2 bear cubs

Children need 1 hat and 2 boots each.

Animals need 1 hat and 4 boots each.

How many hats and boots should Sue make?

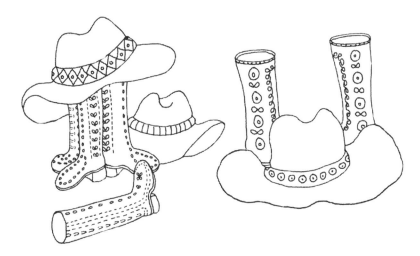

Write Sue a letter.

Tell her how you got the answer.

Pecos Bill, the Greatest Cowboy of All
Cowboy Songs

In this problem, children investigate a given pattern and create one of their own. To begin, children identify Pecos Bill's song-writing pattern, then continue it for seven more days. By counting, adding, and/or multiplying the numbers in the pattern, children can determine that Pecos Bill writes 40 songs in 15 days.

Pecos Bill loved singing cowboy songs around the campfire. It reminded him of his coyote days when he sat on a mountain top and howled at the moon.

Why Pecos even wrote his own songs. Yes, indeed. He wrote lots and lots of them. Here is his song-writing schedule for 8 days.

Day 1	Day 2	Day 3	Day 4
2 Songs	3 Songs	2 Songs	4 Songs

Day 5	Day 6	Day 7	Day 8
2 Songs	3 Songs	2 songs	4 songs

If Pecos Bill follows the same song-writing schedule for 7 more days, how many songs will he write in those 15 days?

Invent another 15-day schedule for Pecos. Make sure it has a pattern. If Pecos follows your schedule, how many songs will he write?

Pecos Bill, the Greatest Cowboy of All
Cowboy Songs

Here is Pecos Bill's song-writing plan for 8 days.
It follows a pattern.

Day 1	Day 2	Day 3	Day 4
2 Songs	3 Songs	2 Songs	4 Songs

Day 5	Day 6	Day 7	Day 8
2 Songs	3 Songs	2 Songs	4 Songs

If he follows the pattern for 7 more days, how many songs will he write in all?

Make a new 15-day song plan for Pecos.

If he follows your plan, how many songs will he write?

Pecos Bill, the Greatest Cowboy of All

Whistling for Snakes

To figure out that Pecos Bill needs 18 rattlesnakes for his lasso, children count by 2s as well as repeatedly add or subtract 2. Changing the story and the quest can extend the problem. For example, when Pecos whistles, have two snakes, instead of one, come to his call. Invite children to find out how many whistles it will take to make a 36-meter snake.

One day Rattlesnake came to Pecos and said, "It is hard work being your lasso, and I am exhausted. Please, give me a vacation."

"But I must have a lasso," said Pecos. "And you are the biggest rattlesnake in all of Texas."

"I know. I don't like to brag, but I am 36 meters long. There has never been and never will be another rattlesnake like me. Fact is, all the other Texas rattlers are 2 itty bitty meters long," said Rattlesnake.

"Don't worry, though," the snake continued. "I have a solution to your problem. Tie a lot of 2-meter rattlesnakes together. Soon you will have a 36-meter lasso and I can rest."

Then the rattlesnake taught Pecos a special whistle. Whenever Pecos whistles in this way, a rattlesnake comes slithering to his call.

Pecos likes the idea of whistling for rattlesnakes. He likes the idea of making a lasso by tying rattlers together. But one thing still bothers Pecos. How many 2-meter snakes does he need to make a 36-meter lasso? Can you help him solve this problem?

Pecos Bill, the Greatest Cowboy of All

The Cyclone

This partitioning problem, which challenges children to separate 100 into five groups, each with a different amount, has many solutions (i.e., 16, 18, 20, 22, 24 or 10, 21, 22, 23, 24 or 10, 17, 18, 25, 30). As they refine their guesses or check their results, children do a great deal of addition and subtraction.

For one long miserable year it did not rain in Texas. Without rain, grass did not grow. Without grass, animals had nothing to eat. The cowboys on Pecos Bill's ranch begged him to bring rain to Texas. So Pecos went all around the country until he found a gigantic cyclone that was full up with rain.

When he found the cyclone, Pecos grabbed his rattlesnake lasso. Then Pecos roped that twisting, wild wind storm. Oh, yes he did. Pecos climbed the cyclone's back and rode it just as though that giant cyclone was a tame little horse.

My, oh my, was that cyclone angry! It twirled and whirled. It bounced and bucked, but it could not get rid of Pecos. Finally, it begged, "Please, Pecos Bill, get off my back!"

"Well now," said Pecos. "I will do just that, but only after we get to Texas."

So the cyclone roared off to Texas, and when it got there, Pecos pulled tighter on the lasso. That tight lasso hurt the cyclone and he began to cry and cry and cry. And do you know what that crying was? Rain. Lots and lots of rain.

The cowboys on the ranch set out buckets to catch all that wonderful rain. When Pecos landed, he counted 100 water-filled buckets in the farmyard. "I guess I'll carry these buckets into the barn," he said.

Pecos made five trips to the barn. He carried a different number of buckets each trip. How many buckets did he carry each trip? Can you figure it out?

Can you think of more ways to solve this problem?

Pecos Bill, the Greatest Cowboy of All

Lots of Lassoing

Children use addition, subtraction, and counting to find three numbers that add up to 15. The number must also satisfy the given clues. Pecos could have only caught 3 bulls, 5 mountain lions, and 7 lizards.

Pecos likes lassoing. It's his favorite way to spend the day. One Monday, he lassoed lizards, mountain lions, and bulls. How many of each animal do you think he roped?

Here are your clues. He caught 15 animals in all. He captured 2 more mountain lions than bulls and 2 more lizards than mountain lions.

Invent your own lasso problem, and give it to a friend to solve.

Lots of Lassoing

Pecos trapped mountain lions, bulls, and lizards with his lasso.

How many of each animal did he trap?
Here are your clues:

- Pecos trapped 15 animals in all.
- He trapped 2 more mountain lions than bulls.
- He trapped 2 more lizards than mountain lions.

Make up your own lasso problem.
Give it to a friend to solve.

Cap O' Rushes

A Tale from England

Long ago and far away, a wealthy man lived with his three daughters in a very grand house. One bright, sunny day this man was watching his daughters play with a puppy.

The man thought to himself, "My children certainly love that little dog. I wonder, do they love me as much? I must find out."

He called his oldest child to his side. "Daughter," he said, "how much do you love your father?"

"Why father," answered the girl, "I love you with all my heart."

"That is good, my dear. That is good," said the man. Then he called his middle daughter. "Child, how much do you love me?" he asked.

"I love you with all my heart and all my thoughts," answered this daughter.

"That is good, my dear. That is very good," said the father. Next he called Fiona, his youngest daughter.

"Little one, how much do you love your father?" he asked.

"I love you as fresh meat loves salt," said Fiona.

This answer enraged the man. He shouted, "What did you say? As much as meat loves salt? I see clearly now. You do not care for me. No, not at all. In that case, I do not want you in my house. Leave, Fiona. Leave today." With these harsh words, the man turned his back on his youngest child.

Although Fiona was sadder than sad, she did not argue with her father. No, she did not. Instead, she selected her three best dresses, a red one, a silver one, and a golden one, and placed them in a sack. Then she walked away from her home.

Fiona wandered in the woods and marched through meadows. She roamed the roads and strode by streams. As she traveled, she gathered rushes. When she had a large armful, she stopped long enough to weave the rushes into a cape with a hood. Wearing this strange outfit, she continued on her way. After a long while, she came to the king's palace. When she arrived at the servants' door, she asked to see the cook.

"If you please," Fiona said to the cook, "can you find work for me in your kitchen? I will do anything."

"Are you willing to scrape and scrub pots and pans until they shimmer and shine?" asked the cook.

"Yes," said Fiona, "I am."

"In that case, the pot-washing job is yours," said the cook.

And so Fiona started working in the palace kitchen. Every day she scrubbed and scraped, from the first golden sunlight of morning until silvery stars dotted the nighttime sky. When her workday ended, she slept in the pantry on a pile of hay.

The cook tried to give Fiona proper clothes, but Fiona refused. She would only wear her covering of rushes. Soon, as a joke, the cook began calling Fiona Cap O' Rushes. Before long, all the servants in the palace copied the cook. In time, no one remembered that Cap O' Rushes had any other name.

One afternoon, the cook burst into the pantry with exciting news. "Tonight, yes, this very night," she said to Cap O' Rushes, "the king is having a dance. All the lords and ladies of the kingdom will be there. Best of all, his majesty has invited the palace servants to come and watch the festivities. Let us go to the party together, Cap O' Rushes."

Did the cook's news excite Fiona? It did not seem to. Indeed, the girl absolutely refused to go to the dance. "I am tired," she said, "much too tired to attend a party. You go. Tomorrow you will tell me everything that happened."

As soon as the cook left the kitchen, though, Cap O' Rushes fetched her sack, the one with her three dresses. She pulled out her red dress, put it on, and then she went to the party.

Cap O' Rushes entered the ballroom all sparkle and flame in her magnificent dress. The prince could not take his eyes off her. He asked Cap O' Rushes to dance. The truth is, he danced with her all night.

The next morning, when the cook entered the kitchen, she said, "Cap O' Rushes, you should have seen the beautiful lady at the party. She captured and held the prince's attention from the first moment until the last. Oh yes she did."

"My," said Cap O' Rushes, "what a sight to see."

"The king is having another dance tonight. Perhaps the beautiful lady will attend. Come with me to the party and see for yourself," said the cook.

"No, no," said Cap O' Rushes, "I am too tired."

That night, as soon as the cook left for the party, Cap O' Rushes got her sack. This time, she withdrew the silver dress, put it on, and went to the dance.

Cap O' Rushes entered the hall all shine and shimmer in her swirling gown. The prince moved to her side and there he stayed all night long.

The next morning, when the cook entered the kitchen, she said, "The beautiful lady was there again last night. The prince did not take his eyes off her."

"Ah," said Cap O' Rushes, "that must have been a sight to see."

"There is another dance tonight. It is the last dance for the year. The prince's lady will surely attend. Come see her for yourself, Cap O' Rushes. I doubt you will ever have another chance," said the cook.

But Cap O' Rushes refused to go.

That night, as soon as she was alone in the kitchen, Cap O' Rushes took out her sack for the third time. Now she withdrew her golden gown. She put it on and went to the dance.

She entered the hall all glimmer and glitter in her glowing gown. She danced and danced and danced, but only with the prince.

Late in the evening, the prince said, "Although I have asked you over and over again, you will not tell me your name. You will not tell me where you live. Promise you will not disappear from my life. I could not stand that."

Then the prince slid a ruby ring on Cap O' Rushes' finger. With that, Cap O' Rushes slipped away from the dance and returned to the kitchen.

The next day the cook told Cap O' Rushes about the beautiful lady in the golden gown. Cap O' Rushes listened and smiled.

Two days went by and then three and four. On the fifth day, the cook came to Cap O' Rushes and said, "I must make some oatmeal for the prince. Poor man, he has not eaten a morsel nor sipped a drink since his lady disappeared from the dance. The queen hopes a bit of oatmeal will tempt him to eat."

"Let me prepare it," said Cap O' Rushes. "I have a special recipe. I know it will help the prince."

Cap O' Rushes fixed the oatmeal and poured it in a bowl. Then, when the cook turned away, she took the ruby ring from its hiding place and dropped it in the middle of the cereal.

The cook carried the bowl to the queen, and the queen handed it to her son. To please his mother, the prince began nibbling the cereal. Suddenly, his spoon hit something hard. What did he hit? The ruby ring, of course. My, he was surprised when he saw that ring.

"Mother," he said, "send for the person who prepared this oatmeal. I want to see her right now."

When the cook arrived, she was afraid to admit that someone else had fixed the food, and so she lied to the prince and the queen. "I made the cereal," she announced. "Yes indeed, that I did."

But the prince shook his head, "You did not make this meal," he said. "Please, tell me who did. Bring her to me."

The cook ran to fetch the true oatmeal chef.

Moments later, Cap O' Rushes entered the prince's room. The queen gasped when she saw a kitchen maid dressed so oddly in a covering of rushes, but the prince smiled. Then Cap O' Rushes took off her cape and stood before the prince and the queen in her golden gown. Now the queen smiled, too. What did the prince do? He asked Cap O' Rushes to be his wife. Did Cap O' Rushes agree? Yes, she did.

That very day, the king and queen began planning a grand wedding. As the wedding day approached, Cap O' Rushes (who was now called Fiona by one and all) asked the queen to invite a wealthy gentleman from a nearby village to the wedding. The queen was happy to fulfill this request. She did not know that the gentleman was Fiona's father. And Fiona's father did not know that the girl about to marry the prince was his youngest daughter.

Fiona did one more thing before the wedding. She went to her old friend the cook and begged a favor. "Please," said Fiona, "for the wedding feast do not use any salt when you prepare the meat."

"I will do what you ask, but I warn you, the food will have a sad, unpleasant taste," said the cook.

"Yes," said Fiona, "I know."

On the wedding night, after Fiona and the prince were married, all the guests sat down for a grand banquet. They began to eat the meat dish, but it was so tasteless no one would have more than a single bite. Then one of the guests began to cry.

"What is the matter, Sir?" asked the queen.

The man replied, "Your Majesty, I had a daughter once. She was my baby girl. I asked her how much she loved me, and she said as much as fresh meat loves salt. I did not understand her words. I forced her to leave my house. Now I know that she did love me after all. But it is too late. She is gone, and I cannot ask her forgiveness. Oh, Your Majesty, I am a very foolish old man."

Just then Princess Fiona stood up. "Father, do not cry. Here I am, Fiona, your daughter who still loves you as much as fresh meat loves salt."

Notes

About the Story

Before telling the story, spend a few moments on a botany lesson. What is a rush? It is a tall plant usually found near marshes. People use the stem to weave baskets, mats, and chair seats. Imagine how uncomfortable it would be to spend all day in a hooded cape made of rushes.

Joseph Jacobs, an English folklorist, originally collected this tale in 1890. I used his version for my retelling.

Jacobs, Joseph. *English Fairy Tales.* New York: Everyman's Library Children's Classics, Knopf, 1993, 57–62.

Your Thoughts

Cap O' Rushes

Meals Without Salt

This counting problem focuses on multiples of 3 up to 100. A calendar will help children to discover that Fiona's father will eat his last salt-free meal at dinner on February 2, and have his first "salty" meal on February 3.

Fiona's father still feels guilty for treating Fiona so harshly. Fiona is not angry at him, but he is very angry at himself. He is so angry he decides to punish himself. How? He will eat 100 saltless meals in a row.

He starts with breakfast on January 1. If he eats 3 meals a day, how many days will he go without salt? What day of the year will he get his first salty food?

Cap O' Rushes

Fiona's New Clothes

To solve this problem, children figure out the number of different three-piece outfits that Fiona can create from two blouses, two skirts, and two hats. As they try to keep track of the different arrangements (permutations), children will come up with different ways to organize and record their work. For additional practice, change the number of items: Add one or more pairs of shoes; remove a hat or two.

Fiona needed new clothes. After all, she could not wear her ball gowns every day. The queen knew just what to do. She sent for the royal blousemaker, the royal skirtmaker, and the royal hatmaker. She ordered them to fashion beautiful clothes for Fiona.

The royal blousemaker made Fiona 2 blouses in 2 different colors: red and white. The royal skirtmaker designed 2 skirts in 2 different colors: blue and yellow. The royal hatmaker produced 2 hats in 2 different colors: green and purple.

On Monday, Fiona wore her red blouse, yellow skirt, and purple hat. On Tuesday, she wore her red blouse, yellow skirt, and green hat. On Wednesday, she created a brand new outfit.

Fiona is amazed at the number of outfits she can create by simply combining her new clothes. Can you figure out the exact number of outfits Fiona can create?

Fiona's New Clothes

Fiona has lots of new clothes.

She has 2 new blouses.
   A red blouse
   A white blouse

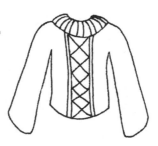 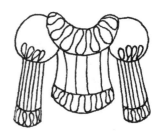

She has 2 new skirts.
   A blue skirt
   A yellow skirt

She has 2 new hats.
   A green hat
   A purple hat

How many different outfits can she create with her new clothes?

Which one is your favorite?

Cap O' Rushes

Everyone Needs Salt

In the first part of this two-part problem, children count by or add halves to find out how many teaspoons of salt a person needs to eat in a given time period. In the second part, children begin by figuring out how many teaspoons of salt are in a salt shaker. They, then must figure out how many days it takes to eat that much salt (at the rate of 1/2 a teaspoon a day). Both parts encourage children to reason proportionally.

Did you know that people cannot live without eating a little bit of salt? It is true. Everyone needs about 1/2 a teaspoon of salt a day. Let's say you do eat 1/2 a teaspoon of salt a day, how many teaspoons will you eat in 2 days?

Can you figure out how much you will eat in 3 days? How about in 4 days? Or 5 days?

How much salt will you eat in 1 week? Can you figure out how much salt you would eat in a month?

Now solve this problem. First, find a salt shaker. How many teaspoons of salt does it take to fill this salt shaker? If you eat 1/2 a teaspoon of salt a day, how many days will it take you to eat a shakerful?

Cap O' Rushes

Everyone Needs Salt

Use the calendar to solve this problem.

If you eat 1/2 a teaspoon of salt a day, how much salt would you eat:

in 2 days?

in 3 days?

in 4 days?

in 5 days?

in 1 week?

in the whole month?

APRIL

Sunday	Monday	Tuesday	Wednesday	Thursday	Friday	Saturday
1	2	3	4	5	6	7
8	9	10	11	12	13	14
15	16	17	18	19	20	21
22	23	24	25	26	27	28
29	30					

Cap O' Rushes

Roses for Pots

Children's first challenge is to sort and decide how to use the given information. The next challenge is to count, add, multiply, or do a combination of these to discover that the prince gives Fiona 525 roses for all the pots she scrubbed.

The prince feels sad whenever he remembers the bad times when Fiona had to scrub pots in the kitchen.

One bright, sunny day the prince woke up with an idea. He planned to give Fiona a single red rose for every pot she scrubbed. The prince knows that Fiona scrubbed 3 pots after every breakfast. She scrubbed 5 pots after every lunch. She scrubbed 7 pots after every dinner.

How many pots did she scrub each day?

How many each week?

**Fiona spent 5 weeks working in the kitchen.
How many flowers will the prince give Fiona?**

Cap O' Rushes

Roses for Pots

The prince wants to give Fiona 1 red rose
for every pot she scrubbed.

Fiona scrubbed 3 pots after every breakfast,
5 pots after every lunch,
and 7 pots after every dinner.

How many pots did Fiona scrub each day?
How many did she scrub each week?

Fiona scrubbed pots for 5 weeks.
How many roses should the prince give Fiona?

The Master Frog

A Tale from Vietnam

Poor Giang Dung (yah-ahng yoom). She had been married only a few months when her kind and gentle husband died. Everyone in the village felt sorry for the young widow. A short time later, though, everyone was amazed. Why? Giang Dung gave birth to a baby. That is not so surprising, I know. But this baby was not a human child. It was a frog. Giang Dung did not know what to think about this odd turn of events. Still, the frog was her child, and she loved him.

Years went by and the frog proved to be a wonderful son. He helped his mother cook the meals, clean the house, and tend her silk worms. In school, he was the best student. After a time, everyone in the village called him Master Frog as a sign of great respect. On his eighteenth birthday, Master Frog made an astonishing announcement. "Mother," he said, "soon I will get married. I plan to marry Kien Tien (kee-en tee-en), the king's youngest daughter."

Giang Dung did not know what to say. True, Master Frog was both kind and smart, but would any girl willingly marry a frog? Would the king's daughter marry a frog? Of course not, it was unimaginable.

"Tomorrow," said Master Frog, "I will tell the king of my plan."

The next day, Master Frog went to the palace and asked to see the king. The guards thought a talking frog would amuse his majesty, so they led Master Frog to the throne room. Master Frog bowed politely to the king before saying, "Greetings, Majesty. My name is Master Frog, and I want to marry your daughter."

When the king heard these words, he roared with laughter. "Well, well, Master Frog," he said between chuckles, "you want to marry one of my three daughters. Which one, I wonder, do you want to wed? Daughters, come here and meet Master Frog. He plans to marry one of you."

The oldest daughter grumbled, "Father, I do not find this frog a bit funny. Send him away immediately."

The middle daughter agreed. "Father," she said, "please remove this arrogant frog from my sight."

The youngest daughter, Kien Tien, said nothing.

"You are right, daughters," said the king. "Enough of this foolishness. Guards, take this animal away and cut off his head."

Before the guards could grab Master Frog, he croaked the loudest frog croak ever heard. At that instant, the palace began to shake. Lightning flashed in the sky. Elephants, tigers, falcons, and panthers came rushing into the throne room. Stamping, growling, pacing, hissing animals surrounded the king, his daughters, the guards, and all of the king's ministers.

Then Master Frog stepped forward and said, "Your Majesty, I command all the beasts who roam the land and fly through the skies. They are my loyal subjects."

The king did not know what to think or what to say. But Kien Tien, the youngest daughter did. She stepped forward and announced, "Father, let me marry Master Frog. In my heart, I know this is meant to be."

The king bowed his head in agreement. And so Master Frog dismissed all the beasts.

Kien Tien and Master Frog were married the next week. It was a grand wedding. As the weeks went by, Kien Tien discovered that she enjoyed her husband's company and depended on his caring support. Indeed, the princess came to love Master Frog with all her heart.

Imagine her distress, then, when she woke one morning to see her husband lying dead on his pillow. Kien Tien cried out in pain and grief. Through her tears, she saw a handsome young man standing by the bed.

"Are you the villain who has killed my husband?" shouted Kien Tien.

"Look in my eyes, Kien Tien," said the man. "You will see I am Master Frog. In your hands, you have the skin I was forced to wear by my father, the Jade Emperor, who is lord of all magic and mystery. I grew up in his fairyland kingdom but I wanted to experience life on the earth. I asked my father to let me come to your world. This request made him angry. He did send me to earth but, in his fury, he insisted I live as a frog. He demanded I stay a frog until I gained the love of a princess. Now that you love me, I can live with you in human form. At least, I can as long as no harm comes to my frog skin. If anything happens to the skin, I must return to fairyland."

Gently, Kien Tien carried the frog skin to a golden cabinet. She put the skin in the top drawer. "It will be safe here, my husband," she said.

Then Kien Tien and Master Frog went to the throne room. Kien Tien introduced her husband to her father and sisters. The king was delighted. "My son-in-law is a son of the Jade Emperor. No one on earth has ever had such an honor," he said.

Kien Tien's sisters, though, were not delighted. No, they were not. They were jealous.

"I am the oldest," one sister complained. "Master Frog should have married me."

The other sister whined and complained. "I think he still looks like a frog," she said.

When the two older sisters were alone, the oldest got an idea. "Let's find Master Frog's skin. Surely, it has magical powers. Perhaps it grants wishes."

The two sisters chose a time when Kien Tien and her husband were away from the palace. Then they entered their room and began hunting for the skin. They looked high and low. They looked here and there. Finally, they looked in the golden cabinet.

They saw the skin. The older sister took it and hid it in her gown.

Late that night, the sisters carried the skin into the garden. They pulled it this way and that. They asked for gold. They begged for diamonds. They requested handsome husbands. Nothing happened. The sisters were furious.

"This is just an ugly frog skin. It has no magic at all," grumbled the older sister.

"Get rid of it. It is disgusting," said the other sister.

The older sister saw a fire burning in the kitchen. She tossed the frog skin into the flames and walked away.

At that very moment, in another room of the palace, Master Frog cried out in pain, "Wife, I am hot, so very, very hot. I feel my arms and legs burning. I cannot breath." And then, with one last gasp, Master Frog died.

Kien Tien could not contain her grief. Her father and sisters heard her agonized screams and ran to her. Through her sobs, Kien Tien described her husband's death.

Immediately, the two sisters realized what had happened. When they threw the skin in the fire, they had killed Master Frog.

"As soon as Kien Tien discovers that the skin is missing, she will know what we did," said the older sister. "And she will tell Father. We must not let that happen."

So the sisters devised an evil plan. They went to Kien Tien and pretended to sympathize with her. They gave her a cup of tea. Was this a nice soothing tea? No, it was not. You see, before giving Kien Tien the tea, the sisters dropped a powerful sleeping potion in the drink.

Minutes after sipping the potion, Kien Tien fell into a deep, deep sleep. Her sisters lifted Kien Tien's unconscious body and carried her away from the palace. They took Kien Tien to a cliff overlooking the sea, threw her into the water, and watched while she sank beneath the waves.

Then the two sisters returned to their father's palace. They were moaning and weeping.

"Oh, Father," wailed the older sister, "our beloved Kien Tien has thrown herself into the sea."

"Yes, Father," said the other sister, "we tried to stop her, but she insisted on joining her husband in the afterlife."

The sisters sobbed and cried, cried and sobbed. At that moment, something amazing, astounding—even miraculous—happened. Master Frog, yes, Master Frog himself, entered the throne room.

He bowed to the king and said, "Majesty, the Jade Emperor, my father, has allowed me to return to your world and finish living my life with your daughter, my beloved wife."

"Dear Brother," said the older sister, "we have tragic news for you. Your wife is dead. She threw herself into the sea. But do not mourn. I will take her place. I will be your wife."

Then the other sister spoke. "Master Frog, if my older sister does not please you, perhaps you would like to marry me?"

Master Frog did not answer the sisters. Instead he raced out of the palace, ran to the cliff, and dove into the sea. He swam under water until he reached a crystal palace. This undersea castle was the home of his uncle, the Dragon King, lord of the waters.

Master Frog approached his uncle and bowed low. As he did, the Dragon King said, "Welcome, Nephew. I have been expecting your arrival. If you turn around, you will see someone else who has been waiting for you."

Master Frog turned. Standing there, more beautiful than a rainbow, was Kien Tien. She smiled at Master Frog and said, "Husband, your uncle rescued me from certain death. He promised you would come for me, and here you are."

Master Frog and Kien Tien left the crystal palace. They swam through the waves until they reached the beach near Kien Tien's home.

It so happened that the sisters were walking along the shore as Kien Tien and Master Frog came up out of the water. Realizing that Master Frog was aware of their evil deeds, the sisters were afraid. Indeed they were so frightened they ran into the forest and were never seen again.

As for Kien Tien and Master Frog, they returned to the palace where they lived happily ever after.

Notes

About the Story

Isn't it amazing how closely this Vietnamese tale resembles the German story *The Frog King, or Iron Henry* recorded by the Brothers Grimm? It seems remarkable that this story line should have a home in two such distant places. You might take this occasion to read the Grimm story to your class. Then pull out a world map and show your students just how far apart Germany and Vietnam are. You could mention that it takes 13 hours and 55 minutes to jet from one country to the other.

These are not the only two frog transformation stories. There is also a Russian tale, *The Frog Princess*, in which the frog turns out to be a magical woman, Elena the Fair. You can find this story in *Russian Fairy Tales* from the collection of Aleksandr Afanas'ev and translated by Norbert Guterman.

Of course, there are many other stories of heroes and heroines forced to live for a time as animals. *Beauty and the Beast* and *Snow-White and Rose-Red* come to mind. Consider sharing these tales with your class. Let your students discuss what it would be like to live as an animal.

Would anyone care to compose a story in which the main character wakes up as a frog? A bear? Or, in Kafkaesque style, a bug?

For non-Vietnamese-speaking Americans, the names in this story may prove difficult to pronounce. If you change the pronunciation from the supplied guide, don't worry. The story will not suffer because you say the name in a unique manner.

For this retelling of *The Master Frog*, I relied on one source.

Vuong, Lynette Dyer. *The Brocaded Slipper and Other Vietnamese Tales*. New York: Harper Trophy, 1982, 65–82.

Your Thoughts

The Master Frog

Paint the House

In this permutation problem, children make different four-color arrangements using each of the colors, red, blue, green, and yellow, just once. Finding several arrangements is rarely a problem. Very challenging, however, is proving that Kien Tien is right: There are 24 ways to paint the house.

What about Master Frog's earth mother, Giang Dung? What happened to her? The fact is Master Frog never forgot this sweet woman. Indeed, with each passing day, he missed her more and more. That is why, with the king's permission, he built her a 4-story house right next to the palace.

Now that the house is built, Master Frog wants to paint it. He doesn't want it to look like any old ordinary house. No, not at all. He wants Giang Dung's house to be the most colorful in the land. So he plans to use red, blue, green, and yellow paint and put a different color on each story of the house. But he cannot decide which colors should go where. Should he paint the first story yellow, the second story blue, the third green, and the top story red? Or should he pick a different arrangement?

How many ways are there to paint the house?

Master Frog thinks there are 12 ways. Kien Tien thinks there are 24. Giang Dung thinks there are 30 ways.

What do you think? How many ways can Master Frog paint the house? Which arrangement do you like the best?

The Master Frog
Jump, Jump, Jump

Partners select several jumping events. For each one, children estimate how many times they will jump, then do the jumping. Besides estimating, counting, and comparing, children get a good physical workout.

Soon Kien Tien gave birth to a daughter, Princess Quynh Dao (quinn yow). The little princess loves hearing about her father's life as a frog. And she loves jumping. She even invented a jumping game. It is also a guessing game. You can play, too.

First you need to pick a jumping event. You could jump while a friend writes the alphabet. You could jump across the room. You could jump while a friend tries to make you laugh. You could jump around a chair.

But here is the important part. Before you start, estimate the number of times you think you will jump during your event. Record your estimate, then test it. Finally record your result.

Now, make up your own events. Write them down, estimate, and then jump, jump, jump.

The Master Frog

Jump, Jump, Jump

Pick an event.

Estimate how many times you think you will jump. Then jump.

Compare your results to your estimate.

- How many times can you jump while a friend writes the alphabet?

- How many times can you jump before a friend makes you laugh?

- How many jumps does it take to jump around a chair?

- How many times can you jump while a friend counts to 100?

Make up your own jumping event.

The Master Frog

How Long Is Your Jump?

Children jump as far as they can, then use a variety of measuring units to quantify the distance. They use non-standard units such as crayons, paper clips, and footsteps as well as a standard unit—the centimeter.

Frogs are famous for being great jumpers. How good a jumper are you? How far can you jump?

Place a piece of masking tape on the floor. Stand on it. Then take your longest jump. Mark the spot you land on with a new piece of tape. Then measure your jump in many different ways.

Measure it with crayons, paper clips, your footsteps, your hands, your thumbs, and with a centimeter-ruler.

The Master Frog
How Long Is Your Jump?

Work with a partner.
Take your longest jump.

How many crayons long is your jump?

How many paper clips long is your jump?

How many footsteps long is your jump?

How many hands long is your jump?

How many thumbs long is your jump?

How many centimeters long is your jump?

Find a new way to measure your jump.

The Master Frog

Frog Facts

After they talk about and solve two different frog problems, children make up their own problems—about frogs, of course. Children choose from a list of frog facts and use their imagination to create situations that use one or more arithmetic operations.

Frogs are remarkable animals. Here is a list of true facts about these jumping creatures. Think about the facts. Then use the information you like best to make up your own math problems.

Here is a problem based on the facts: If one frog can eat 100 mosquitoes in a night, how many mosquitoes can 6 frogs eat in a night?

Here is another problem: A frog can eat 100 mosquitoes a night. But if a frog doesn't happen to be hungry, he might eat 20 mosquitoes less than that. If he did, how many bugs would he eat?

The Master Frog
Frog Facts

Here is a list of frog facts.
Use the facts to make up math problems.

- Frogs have 5 toes on each back foot and 4 toes on each front foot.

- The largest frog is a goliath from Africa. Goliaths can be 14 inches long and weigh 7 pounds.

- The smallest frogs are less than 1/2 inch long.

- Many frogs can jump 20 times their body height. This means a 1-inch frog can jump 20 inches.

- The longest recorded frog jump is more than 20 feet. In 1984, a frog named Weird Harold made this huge leap in a frog-jumping contest.

- A frog can eat 100 mosquitoes in one night.

- A frog can lay 1,000 eggs at one time.

- There are at least 4,100 different kinds of frogs in the world.

- The arrowpoison frog has skin that oozes the most poisonous venom in the world. A single 1-inch frog can kill 50 people. This frog lives in Central and South America.

The Silk Worms

As they try to find out how many leaves Quynh Dao collects each week, children count, add, and double numbers. All the numbers in this problem, except for 5 and the solution, 315, are multiples of 10.

Giang Dung loves spending time with her granddaughter, Quynh Dao. Together they care for Giang Dung's silkworms. This is not an easy job. You see, silkworms eat mulberry leaves and nothing buy mulberry leaves And, even though the worms are tiny, they eat an enormous amount of food. Every day Quynh Dao gathers leaves and feeds them to the worms. Since the princess likes doubling things, she invented a doubling schedule for her leaf collections.

On Monday, the princess picks 5 leaves. She doubles that amount on Tuesday. On Wednesday, she doubles Tuesday's leaves. On Thursday she doubles Wednesday's total. On Friday, she doubles again. And then she doubles one more time on Saturday.

She never picks leaves on Sunday. Can you figure out how many leaves Quynh Dao collects each week?

A Tale from Germany

There was once a wealthy gentleman who had three sons. Everyone agreed that the two oldest were charming and clever, whereas, to one and all, the youngest seemed a strange and foolish lad. Unlike his brothers, he never socialized with the elegant guests who visited his father's home. Instead, he spent his time in the garden talking to birds, playing tag with rabbits, and taking care of his rose bushes. And so, people called the youngest son Simpleton.

One day, the two oldest sons decided to leave their father's home and seek their fortunes in the wide, wide world. Days, then weeks passed, and the brothers never returned. Finally, Simpleton went to his father and said, "Tomorrow, with your permission, Father, I will go out into the world and start searching for my brothers. I will find them and bring them home."

At first the gentleman refused to grant Simpleton's request. Why? He was afraid his youngest son, being such a simple soul, would come to harm during his search. Simpleton did not give up, though. He begged and begged until his father gave his blessing.

As it turns out, Simpleton did not search for long. The truth is he discovered his brothers in a nearby town, living at the nicest inn. Here they spent every day in the same way: eating, drinking, and not paying their bills. When Simpleton appeared at the inn door, his brothers were surprised and delighted. Not because they liked Simpleton, oh no, but because they knew their kindhearted brother would pay all of their debts. Simpleton did pay all that was owed. Then he and his brothers left the inn, but they did not return to their father's home. Instead the three brothers set out together to seek their fortunes.

They had not traveled far when they saw an anthill by the side of the road. The oldest brother said, "Come on, let's have some fun. Let's make these bugs hurry and scurry. Then he seized a stick and started to poke at the anthill.

"This is a funny sight!" chuckled the second brother.

But Simpleton grabbed the stick from his brother's hand. "Leave these innocent creatures alone," he demanded. "I will not allow you to hurt them." The older brothers laughed at Simpleton's foolishness, but they did not disturb the ants any further.

Soon the brothers came to a lake. There they saw a family of ducks paddling about the water. "I'm hungry," said the oldest brother. "Let's catch these ducks and cook them for dinner."

The second brother agreed. "A duck dinner," he said, "how delicious."

But then Simpleton spoke up, "No, Brothers, you must not hurt these gentle birds. Let them swim in peace."

The brothers made fun of Simpleton's sympathetic nature, but they did not hurt the ducks. Instead, they continued on their way. Soon they came to a big tree. On the tree limb they saw a huge bees' nest. This nest was positively spilling over with sweet, golden honey. The oldest brother said, "Let's make a fire under this nest. The smoke will suffocate the bees. Once they are dead, we can take all the honey we want."

Before the second brother could strike a match, though, Simpleton said, "Brothers, do not kill these little ones. They have not harmed you. Let them stay safe in their home."

"Simpleton," grumbled the oldest brother, "it is getting dark and thanks to you, we have nothing to eat. Why, we do not even have a place to sleep."

Simpleton did not answer his brother. Instead, he pointed straight ahead. Through the trees, in the deepest part of the forest, Simpleton pointed to a large castle. The brothers walked and walked until they reached this woodland palace. When they arrived, they discovered that the castle was a very strange place indeed. First they entered the stables. These stables were full of horses, as you might expect, but the horses were all made of stone.

Next the brothers walked into the main hall. Here they saw dozens and dozens of stone statues, only these statues were all of people. Aside from the statues, the room was completely empty. The brothers continued walking through the castle without seeing a living soul—not a servant, not a lady or gentleman, not a child.

Finally, at the top of the tallest tower, they came to a room with a locked door. The door had a small window in it. When the brothers looked through that window, they saw a gray-haired man sitting at a table. They called to him. They called once. They called twice. But the old man did not respond. On their third try, though, the man raised his head. He got up from his table and unlocked the door. Without speaking a word, he led the three brothers through long, winding corridors until, at last, they came to a banquet hall. There they saw a table covered with meats, breads, cheeses, and pies.

After the brothers finished eating a huge dinner, the old man pointed toward a stone table in the center of the room. When the brothers looked closely, they could see that someone had carved words into the stone. These words explained the mystery of this peculiar castle.

You see, the castle and all its inhabitants, including the king and his three daughters, were under an evil enchantment cast many, many years ago by a cruel, jealous witch.

To break the spell, some valiant hero must perform three impossible tasks, and he must perform them in a single day between sunrise and sunset. For his first task, this hero has to find and gather together the 1,000 perfect pearls that make up the youngest princess' favorite necklace. On the day the witch cast her spell, she scattered these pearls throughout the castle garden. If the hero fails to find the pearls, if even one precious jewel is missing, the hero will be turned into stone.

When the oldest brother read these words, he said, "It sounds easy enough to break this spell. Tomorrow morning, I will collect the pearls."

At sunrise, the oldest brother walked into the garden. All day, he searched. He looked under bushes and in clumps of moss. At sunset, he had 100 pearls in his hand. As the last golden rays of sunlight faded in the sky, the oldest brother turned into a statue of cold, hard stone.

That night, the second brother said, "Tomorrow I will hunt for the pearls. If I succeed, perhaps I will free our brother from this evil enchantment."

The following morning, the second brother began searching. When the sun set, he had 150 pearls. And so, this young man, too, became a creature of stone.

On the third morning, Simpleton took his turn. He searched and searched, but at noon, he had just 200 jewels. "At this rate," he sighed, "I will never find all the pearls in time." Just then he saw something amazing. Ants. Everywhere Simpleton looked, he saw ants.

Suddenly, the largest one started talking. "Dear friend," she said, "let me introduce myself. I am the queen of these ants. Three days ago, you saved our home and our lives. Please let us show our gratitude by performing a service for you. We can find the pearls."

As Simpleton watched, the ants spread out into every corner of the garden. In a few minutes, they returned, and Simpleton saw that each little ant was pushing a great big pearl through the grass. Simpleton counted. Sure enough, he had 1,000 pearls. He placed them in a leather pouch and put the pouch around his neck. Of course, Simpleton had not broken the spell yet. He still had two more tasks to perform before the day's end.

The second task was just as hard as the first. Simpleton had to find the golden key that unlocks the bedroom door where the three princesses slept. The witch had tossed this key into the deepest, muddiest part of the castle lake.

When Simpleton reached the lake, can you guess what he saw? He saw the ducks whose lives he had saved. Before Simpleton could say a word, the largest duck swam to shore. He held the golden key in his mouth. As he dropped the key at Simpleton's feet, the duck said, "Thank you, kind sir, for saving our lives. We will be grateful to you forever." Then the duck family swam away.

With the pearls in his pouch and the key in his pocket, Simpleton was ready for the next task. He walked to the princesses' bedroom and unlocked the door. Inside, he saw three beds and on each bed was a princess. Now, the odd thing was that each princess looked exactly, identically, precisely like the other two. It seemed impossible to tell them apart. And yet, that was the third task. Simpleton had to go to the youngest princess and place the pearls on her bed. But how could he tell which princess was the youngest? There was only one difference between the three. You see, before being enchanted, the oldest princess had eaten a bit of sugar, while the second sister had tasted a spoonful of syrup, and the youngest princess had enjoyed a helping of honey. Simpleton stood staring hopelessly at the three sleeping beauties, wondering how he could ever select the youngest. Then a bee, a queen bee, flew through the window.

The bee landed on Simpleton's shoulder and whispered in his ear, "Thank you, sweet friend, for saving my nest. Now, allow me to do a service for you."

The queen bee buzzed over the three princesses before landing softly, delicately, on the mouth of the youngest, the one that had eaten honey before falling into her enchanted sleep. Simpleton took the pearls and placed them on her bed. The moment he did so, the princess awakened. So did her sisters and her father, the king. And then all the statues, including those of Simpleton's brothers, turned back into living, breathing men, women, children, and horses.

What happened next? There was a great celebration that lasted for weeks. When the festivities ended, the youngest princess declared her love for Simpleton, and Simpleton knew that he loved her, too. And so, they got married. Then, the old king died, and Simpleton was made ruler of all the land. What about his brothers? They married the two older princesses and lived with Simpleton in comfort and happiness ever after.

Notes

About the Story

Ants and bees are fascinating and, in many ways, very similar insects. They both live in colonies dominated by queens. Every member of the colony has a particular job. They are remarkably industrious. This story might present a good opportunity to research these extraordinary creatures.

For this retelling I used two versions of the story.

Grimm, Jacob and Wilhelm. Crane, Lucy, trans. *Household Stories: From the Collection of the Brothers Grimm*. New York: Dover, 1963, 262–264.

Grimm, Jacob and Wilhelm. *The Complete Grimm's Fairy Tales*. New York: Pantheon Books, 1972, 317–319.

Your Thoughts

The Queen Bee

Sleeping Ducks

As they consider the clues given in this logic problem, children use deductive reasoning to decide where each duck sleeps. Cubes or markers are especially helpful in arriving at the solution.

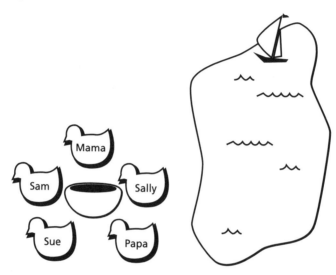

Simpleton invited the duck family to live by the castle lake. There are five ducks in the family: Mama, Papa, and three ducklings named Sam, Sally, and Sue. Mama and Papa build a nest in the grass near the water. The nest is a great big circle. Each duck has his or her own special sleeping spot in the nest.

Can you figure out where each duck sleeps?

Here are your clues:

- Mama sleeps at the top of the nest.
- Mama sleeps between Sam and Sally.
- Papa sleeps between two girls.
- Sam sleeps next to Sue.
- Papa sleeps near the lake.

The Queen Bee
Sleeping Ducks

Where do they sleep?

Here are the clues:
- **Mama sleeps at the top of the nest.**
- **Mama sleeps between Sam and Sally.**
- **Papa sleeps between two girls.**
- **Sam sleeps next to Sue.**
- **Papa sleeps near the lake.**

Write a name on each duck.

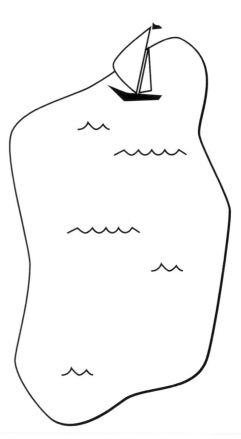

The Queen Bee

Lots of Honey–Version 1

This version—the easier of the two—gives children the opportunity to reason proportionally with halves. Since one honey cake requires half a jar of honey, two cakes require one jar. Thus, the cook can make two cakes a week and 104 in a year

The queen bee and all her followers decided to create a new hive in the castle garden. Simpleton was happy to have his friends so close at hand. He was even happier when the queen bee announced that her workers would deliver 1 jar of honey every week to the castle cook. Why? Simpleton loves to eat honey cakes.

It takes 1/2 jar of honey to prepare each cake. Can you figure out how many cakes Simpleton will get each week from 1 jar of honey?

Then think about the 52 weeks that make one year. How many cakes will Simpleton have in a whole year?

The Queen Bee
Lots of Honey–Version 1

The queen bee gives 1 jar of honey to the
castle cook each week.
The cook uses the honey to make honey cakes
for Simpleton.

It takes 1/2 jar of honey to make a cake.
How many cakes does the cook
make each week?

There are 52 weeks in a year.
How many cakes does the cook
make in a year?

Lots of Honey–Version 2

In this version, children again reason proportionally, but this time, with fourths as well as halves. After using 1/4 of a jar of honey for the tarts and another 1/4 for the castle tea, the cook has 2-1/2 jars left—enough to make five cakes. Thus, Simpleton can eat 260 honey cakes in one year.

The queen bee and all her followers decided to create a new hive in the castle garden. Simpleton was happy to have his friends so close at hand. He was even happier when the queen bee announced that her workers would deliver 3 jars of honey every week to the castle cook. Why? Simpleton loves to eat honey cakes.

The cook uses 1/4 jar of honey each week to sweeten the castle tea. She uses 1/4 jar of honey each week to bake tarts for the queen. She uses the rest of the honey to prepare honey cakes. Her recipe calls for a 1/2 jar of honey for each cake.

Can you figure out how many cakes Simpleton gets to eat each week?

There are 52 weeks in a year. Now can you figure out how many cakes Simpleton eats during the year?

The Queen Bee
Lots of Honey—Version 2

The queen bee gives 3 jars of honey to the castle cook each week.

The cook uses 1/4 jar to sweeten the tea. She uses 1/4 jar to make tarts for the queen.

The cook uses the rest of the jar to make honey cakes for Simpleton.

It takes 1/2 jar to make each cake.
How many cakes does the cook make each week?

There are 52 weeks in a year.
How many cakes does the cook make in a year?

The Queen Bee

A Place for Animals

This is a two-part problem. First, children partition 35
into five equal groups to find out how many of each
different animal Simpleton counts. Then, to find out
what happens the next day, children add and subtract
the given information from 7 (the solution to part one).
That day, there are 34 animals—4 deer, 11 rabbits,
9 bears, 5 wolves, and 5 foxes.

Before long, word spread throughout the forest: Simpleton,
the new king, wanted to protect all the animals. If an animal
found its way to the castle garden, the creature would find
peace and safety. Hunters knew better than to hurt Simpleton's
animals. The animals rarely stayed on the castle grounds,
however. They were wild animals, after all, and they needed to
run free in the forest.

One day Simpleton counted 35 animals visiting his garden
sanctuary. There were deer, rabbits, bears, wolves, and foxes.
There were the same number of each animal. Can you figure
out how many deer, rabbits, bears, wolves, and foxes were
staying in the castle garden?

The next day, 3 deer left the garden to run in the woods.
Two bears came into the garden and 3 foxes left. One fox
changed his mind and came back. Two wolves left and 4 rabbits
came. How many animals are in the garden now?

The Queen Bee
A Place for Animals

One day there were 35 animals in the garden.
There were deer, bears, foxes, wolves, and rabbits.

How many of each animal were in the garden?

The next day
3 deer left,
2 bears came,
3 foxes left, but 1 came back,
2 wolves left, and
4 rabbits came.

How many of each animal are in the garden now?
How many animals are there altogether?

The Queen Bee
Designed by Ants

In spatial visualization problem, children connect pairs of points with straight lines, then count the triangles they see. Demonstrate Simpleton's favorite design on an overhead projector using a transparency of page 124. Because some triangles in this design overlap, the number of triangles that children count will vary from 6 to 10. The easiest triangles to spot are A, B, C, D, and E. A combination of these triangles gives four more: B + C, E + F, B E, and C + F.

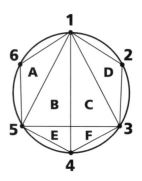

Simpleton was grateful to the queen ant and all of her followers. So he invited the entire colony to live in his castle garden. The ants loved their new home. They played a special game on the castle grounds. First the queen made a big circle in the dirt. How did she do it? She walked round and round and round. Then the ants ran in straight lines from one part of the circle to another. When they were done, the ants stood back and looked at their beautiful design.

The ants made a new design every day. And every day, Simpleton went to the garden to see their work. Here is one of Simpleton's favorite designs.

To make it, you start with a circle that has 6 points around its edge. Now draw lines from point to point. Connect 1 to 3, 3 to 5, 5 to 1, 1 to 2, 2 to 3, 3 to 4, 4 to 5, 5 to 6, 6 to 1, and 1 to 4.

Study the design you made. How many triangles are hiding in it?

Now, make your own designs. Then count the triangles.

The Queen Bee

Designed by Ants

Here are circles. Draw a design in each one.
Go from point to point.
Now, count the triangles you see.

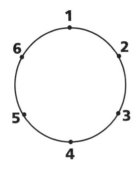

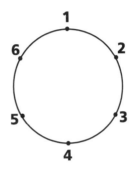

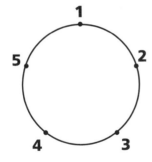

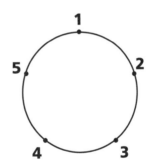

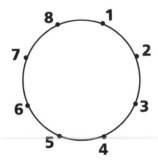

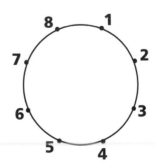

The Queen Bee
Designed by Ants

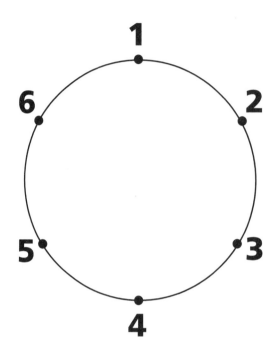

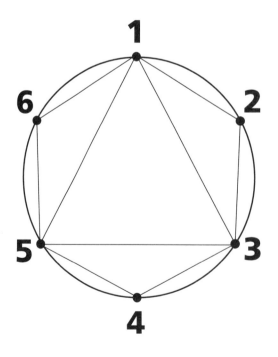